JUDGEMENTS OF VALUE

Judgements of Value

Selected Writings on Music

by
Martin Cooper

Edited by
Dominic Cooper

Oxford New York
OXFORD UNIVERSITY PRESS
1988

Oxford University Press, Walton Street, Oxford OX2 6DP
Oxford New York Toronto
Delhi Bombay Calcutta Madras Karachi
Petaling Jaya Singapore Hong Kong Tokyo
Nairobi Dar es Salaam Cape Town
Melbourne Auckland
and associated companies in
Berlin Ibadan

Oxford is a trade mark of Oxford University Press

Published in the United States
by Oxford University Press, New York

British Library Cataloguing in Publication Data
Cooper, Martin, 1910–1986
Judgements of value: selected writings
on music.
1. Music. Criticism
I. Title. II. Cooper, Dominic, 1944–
780'.1'5
ISBN 0–19–311929–3

Library of Congress Cataloging in Publication Data
Cooper, Martin, 1910–
[Selections. 1988]
Judgements of value: selected writings on music/by Martin
Cooper: edited by Dominic Cooper.
Includes index.
1. Music—History and criticism. I. Cooper, Dominic, 1944–
II. Title.
ML60.C822   1988   780'.9—dc 19   88–9427
ISBN 0–19–311929–3

Set by Cambrian Typesetters
Printed in Great Britain
at the University Printing House, Oxford
by David Stanford
Printer to the University

FOREWORD

by Sir Isaiah Berlin

Martin Cooper was one of my oldest friends. From the beginning there was a natural bond of sympathy between us which never grew weaker, and made communication between us, at all times, easy and usually exhilarating. I remember our first meeting: it was in the spring or summer of 1930 (or 1931) in the ante-room of the Holywell Music Room in Oxford, after a recital of his own piano works by the now largely forgotten Russian composer Nikolai Medtner. We walked away together and discussed the quality of Medtner's music, the possibility of the influence on him of Brahms, Tchaikovsky, Scriabin, the social views of Russian composers at the beginning of the twentieth century (I remember that one of Medtner's works was called *Ode au travail*, which seemed to indicate vaguely socialist sympathies). Then we talked about contemporary English composers, about whom Martin tended to be moderately critical. I thought I have never before met so spontaneous and fascinating a talker with so wide and lively an interest in the European art and culture of our time. We agreed to meet again, and so began a happy personal relationship which lasted to the end of his life. His friends, I discovered, were some of the most original and celebrated figures in Oxford—Maurice Bowra, John Sparrow, Goronwy Rees, A. J. Ayer, Alan Pryce-Jones, Duff Dunbar. Martin had been educated at Winchester, but there were not, it seemed to me, many Wykehamists among his friends—John Sparrow was one, but apart from him I knew of no schoolfellow of his who could be described as an intimate friend.

One spring vacation—it may have been in 1931—he invited me to stay with him at his parents' house in York. His father was a Canon at the Minster, a charming, gentle, infinitely welcoming host; his mother a rather more formidable and firm figure, who seemed to me to dominate the household. I stayed for about a week. In the evenings I tried to read philosophical books as part of my academic course. The old Canon would come up to the chair in front of the fire in which I was sitting, look over my shoulder, look at my worried expression, and ask 'Stiff?' 'Yes,' I would answer, 'very'. He would shake his head in sympathy and pat me gently on the shoulder. Martin and I went for walks in York and its environment: I remember that he took me to a tea-shop near the Close, which entertained its customers with some kind of forerunner of the juke-box—when I put the relevant coin in it, it played the 'Dance of the Seven Veils' from Richard Strauss's *Salome*. I don't know why, but I thought it curiously incongruous in those clerical surroundings, for all its Biblical associations. Martin thought it would do the pious nothing but good.

In Oxford I saw a great deal of Martin—he had plenty of time to spare: he was not a particularly assiduous student, and, after failing in some examination, was requested to leave Hertford College, of which he was an undergraduate, and he migrated to St Edmund Hall, which was more tolerant of lively academic dilettanti. His company was a source of endless pleasure to me: he was spontaneous, affectionate, imaginative, gifted, amusing, with an acute and often ironical interest in the characters or peculiarities of human beings, in literature, in the arts, in, so it seemed to me, every aspect of the life he led, or that of writers and composers in whom he took an interest. He was infinitely gay and responsive to both ideas and works of art, reacted to them at once more impulsively and more intelligently than most of our common friends. After he took his degree I did not see him for a year or two—I knew that he had gone to Vienna and become a student of the head of the Conservatory in that city, the celebrated teacher and composer Egon Wellesz. When Wellesz, who was by origin a Jew, was driven out of Vienna in the later thirties, Martin had a large part in getting him domiciled in England—Wellesz spent most of his remaining years, contentedly enough, I thought, as a greatly respected Fellow in music of Lincoln College, Oxford. Martin remained attached to him, and learnt a great deal from him about out-of-the-way subjects in which he delighted—the music of the early Ethiopian and Armenian churches, the influence of Venice upon the Armenian church in that city and vice versa; the development of atonalism in Vienna and its musical politics and its impact on composers; and in addition to this, much solid learning. I remember meeting Martin at the Salzburg Festival in, I think, 1934, armed with sheets of music paper on which, he told me, he proposed to inscribe four songs which he had composed, somewhat in the style of Schumann, but very different from him too. At that time he wished to be a composer and a pianist, and it was only after he had convinced himself that his work in both these spheres did not come up to his unshakeable critical standards that he abandoned this for the history and criticism of the work of others.

I saw him on and off in London in the thirties—he came to stay in my parents' house once or twice, and, as before, we talked about everything in the world. One of his remarkable attributes was his understanding, both intellectual and intuitive, of what a culture is, and what the impact of one culture on another could generate. He understood, and could discuss vividly and with accurate knowledge, the impalpable relationships of art—music, painting, architecture—to social outlooks, ways of life, philosophical trends, and above all, the literature, both imaginative and critical, of a given form of civilization; and, of course, of religion—its ritual, its institutional life, inner spirit, mythology, tradition, which had entered so deeply and inevitably into the European consciousness and creative activity in every sphere, from the earliest days. His deep and sensitive grasp of the strands that bind together different aspects of the life of a given society at some particular stage of its development was something that not many musical critics of his time, at any rate in England, have, or want to have. It was this that enabled him to say and write such illuminating things about, say, the connections of the opera to life of the seventeenth- or eighteenth-century courts, to the literary forms chosen by the librettists, to the effects of patronage on artists, of the influence of movements of social revolt, or bold, new intellectual ideas on the music composed by the creators of the operatic tradition. I had never heard these topics more brilliantly and attractively expounded than in his conversation directed at increasing my understanding of these things.

Even then—in the mid-thirties—it was plain that what attracted him most deeply was the music (and literatures) of Latin countries, particularly France and Italy. His interest in

Eastern Europe, especially Russia, developed somewhat later. Martin was, of course, steeped in the German musical tradition (as what writer about music cannot afford to be), but French eighteenth-century opera fascinated him more than anyone else I know. It was from his lips that I heard such names as Philidor, Monsigny (and much about them), and not only those of Rameau or Gluck; and then of Paër and Méhul (who else at this time could have discoursed on his unperformed *Agar au désert*, or compared *La Journée aux aventures* to Laqueur's *La Mort d'Adam et son apothéose?*). It was all far more interesting than anything by Romain Rolland. He adored Berlioz, when not so very many people in England took an interest in him and spoke of him as a composer whose music possessed magical qualities. He entered realms not traversed by others. He loved *opéra comique* as such, on which he wrote, as we all know, excellent studies, both in books and in articles. He adored Verdi, too, and once said to me that he would come from the depths of China if there were a Verdi festival anywhere in the world, whatever the distance. Needless to say, he profoundly admired the great German masters (his book on Beethoven is evidence enough of that), although he remained somewhat cool to Wagner, and to Schoenberg and serialism. I did not know anyone else in England who spoke so eloquently and, to me, convincingly, of the beauties of Boieldieu, Auber, Meyerbeer, and of course, Gounod, but above all, Bizet (on whom he wrote an excellent monograph), as Martin. He loved Fauré, Debussy, Ravel—the French 'silver age'—and the minor deities too—Chausson, Roussel, d'Indy. He shared my distaste for César Franck ('the salacity of the organ loft', he once said of him), and Florent Schmitt. He did his duty to British composers in the British Council booklet—and said that if Vaughan Williams had been called Vagano-Guglielmetti he might have had a better press in Europe. Although his musical heart remained in France (when it did not wander to Russia), he thought that the new generation of British composers was seriously underestimated in that country.

I did not see a great deal of him in the immediate pre-war years. I read his musical articles in the old *London Mercury*, the *Daily Herald*, the *Spectator*, and we exchanged occasional letters. During the war I was in Washington and Moscow on government service, and it was only after I returned, in 1946, that he told me that he had been received into the Roman Catholic Church in the first year of the war. I remember telling him that the Thomist philosopher Jacques Maritain had once told me that he was converted because he found himself on a path which led unswervingly towards the Church, that he felt no sense of relief by becoming a Catholic, but that he could not do otherwise; Martin said that this was not dissimilar to his own condition, but plainly did not wish to discuss this further, at any rate with me. It seems to be unforgivably arrogant to pronounce on the source, or, indeed, nature of another human being's spiritual experience, however well one may believe that one understands it. I go no further than to advance, most tentatively, that Martin's deep understanding of the Western cultural tradition, especially in France and Italy, in which the ritual, institutions, faith, moral and metaphysical ideas of the Roman Church played so central a part, had something to do with his conversion—maybe in reaction to the somewhat chilly Protestantism of the north of England, into which he had been born, and from which he recoiled aesthetically and ethically too. His conversion lost him some of his militantly anti-clerical friends, but our friendship remained undisturbed.

One of the links between us—not that we needed it—was our common friendship and admiration for Miss Anna Kallin, a most remarkable woman, who from the beginning had shaped and developed the talks department of the new Third Programme of the BBC, encouraged by her Director, and life-long admirer, Sir William Haley. Miss Kallin was Russian, brought up in Moscow, and had spent many years in Berlin. She was a many-sided, highly intelligent woman, with a sharp sense of humour and immense culture, lively, amusing and amused, with an exceptional gift for discovering and eliciting broadcast talks from some of the most talented members of the young post-war generation. To see Martin and her together was a great pleasure: intellectual gossip, endless dissection—sometimes mocking, at other times affectionate and admiring—of the personalities and motives of others, free play of irrepressible ridicule of anything that seemed pompous or pretentious or silly or philistine (there had been a good deal of opposition from such quarters to Miss Kallin's unyielding standards), made their association a source of delight to their friends and themselves.

At some point in his later life Martin's religious orthodoxy melted away. I have learnt from his daughter Imogen that in his desk were found passages of poetry—free translations from the Italian of Leopardi, the Spanish of Machado, the French of the Romanian Petru Dimitru—which had meant much to him in his last years. They have in common a noble despair, a painful acquiescence in the face of vast, irresistible, unintelligible forces, a longing for the dark path which leads into the night, to *le néant*—a welcome guest, as in Lensky's aria before his death in Tchaikovsky's *Eugene Onegin*, or Heine's poem set to music by Brahms:

> Der Tod das ist die kühle Nacht
> Das Leben ist der schwüle Tag.

The Russian poem, much loved by him, which Martin asked me to read at Miss Kallin's funeral has some bearing on this. They might both have agreed with the Russian writer Alexander Herzen, who said that art, which resists decay, and the summer lightning of happy love, are all that we can cling to in our lives, which apparently lack all meaning and purpose.

In our relationship he continued to the end to talk with unique charm and animation about people and things. His critical judgement, his wit, his generosity of spirit, were undimmed. His dark metaphysical questionings did not, it seemed to me, diminish the sensibility with which he responded, often passionately, to anything that seemed to him to possess artistic vitality, to the great monuments created by human genius, to anything that resisted the forces of barbarism and destruction.

His friendship was one of the great blessings of my life, and I shall never cease to mourn his passing.

Reprinted from Sir Isaiah Berlin's tribute at the memorial concert, 29 June 1986

PREFACE

I have always thought that my father's description of the music critic as being, by needs, an 'amphibian', at home in two worlds at once, concealed a gleam of perspicacity about himself: for within his own personality there were a number of such divisions and contradictions. These, I think, he regarded as something unresolved and therefore in the nature of a weakness; but seen more dispassionately than his rigorous self-criticism would allow, these differences and opposing viewpoints seem to me to have played a large part in giving him such an all-round appreciation not just of music but of the whole world of art and culture, and history which he so loved.

Of the various disparities in him, the most crucial was that between his emotions and his intellect—and nowhere did this more clearly show than in his inner life. With six consecutive generations of Anglican clergymen on his father's side (and several more on his mother's), his heritage was patently not one to be shrugged off with any ease; and in the event, though he turned things about by converting to Catholicism in his late twenties, he nevertheless remained essentially within the family tradition in that his private, spiritual life was always his predominant concern.

His reaction, in fact, seems to have been largely against the parochial in his Anglican upbringing: what he turned for was perhaps religion on a grander scale, a Latin church with its mysteries and ritual whose involvement at every level of civilization over the centuries had made it the subsoil on which European art had grown. Here he found faith; and it was this, with all its emotional force, that gave him those absolutes—without which he said he found life both ugly and pointless—which his inborn tendency to intellectual scepticism seemed otherwise to have denied him. 'A real doubting Thomas', his father had called him; and while this instinct to question stood him in such good stead as a critic, it was also what caused him most discomfort within himself.

Although his interests had always been centred round things

of the intellect, he wore his learning lightly, talking about such matters totally naturally and never making others aware that their knowledge might be anything less than his. But to talk to him was always to learn. Yet at the same time he had a genuine horror of being thought 'an intellectual'; and perhaps because of this and also to ward off the pomposity that can so easily creep into intellectual discussions, he was always full of wit and that wry, throw-away humour (as often as not aimed at himself) which was one of his chief characteristics.

He responded to both people and ideas with equal spontaneity and warmth; and was generous in his judgements towards anything that he believed to be sincere—yet scathing of things that smelt to him of pretentiousness or humbug. Unlike some of his colleagues, however, he did not think it either helpful or right to be merely perniciously damning in print. For him, adverse criticism had to be constructive if it were not just to lapse into prejudice and journalistic showmanship. He never forgot just how precarious is the position of a critic of the arts.

If it was that edge of scepticism tempering his more impulsive feelings which made such a good foundation for his critical judgement; and if it was the same intellectual sharpness that ran parallel to and balanced his faith, there were other divisions in him too. The man of intellect that the public saw also had a side of strange simplicity and *naïveté*; the modest man (and he was quite free of any pretensions) was dogged by a private pride that had its source in his own self-doubt. But over and above these, the overall image that he has left behind is that of a man who was essentially good: good in that, for all his concern with his own inner state, he lived as if those around him—and indeed anyone in need—were more important to him than himself.

He had his blind spots of course—such as organ music (a reaction to his church upbringing no doubt), Mahler (whose emotionalism coincided too uncomfortably with his own), detective films (these he thought just 'silly'), and anything remotely mechanical (he never learned to drive nor got to the finer points of using a typewriter)—and I think that the severest of these, at least for his own peace of mind, was his inability to accept that the truth was ultimately beyond knowing. In the end, this lifelong search and all the questioning that went with it—his own, private *advocatus diaboli*—gained the upper hand

and swept away his earlier sureness. Yet for all this, he said, little really changed. 'What I had lost was better lost,' he wrote in 1983. 'It had to make room for an old man's vision of the "truth", that we are always trying, and always failing, to grasp and tie down. Yet there is absolutely no break with the past: my fundamental ideals (and indeed beliefs) have remained virtually unchanged.'

One of the things that was strongest in him and gave him most pleasure was his sense of the comic and ridiculous in all of us. When something of this struck him, his laughter was uncontrollable and infectious—as if to behave (even if only momentarily) as though the whole of life were just a joke was an enormous relief to him. A part of him, I believe, longed to be free of the seriousness and ever-present spiritual awareness that he had inherited from those clerical ancestors of his. He said, in fact, that if he were remembered, he hoped that it would be more as a joker than as a preacher. . . . Shortly after his death, we found some notes that he had made about what he would like done with this and that small possession—and a few suggestions and preferences about his burial. 'But don't feel bound to follow these,' he ended. 'After all, it's your funeral.' Always one for a *bon mot*, it was typical of him that he should leave this valedictory quip to ensure that even at his own death there was room for a good laugh.

Dominic Cooper
Achateny, January 1988

ACKNOWLEDGEMENTS

The articles in the section entitled 'The World of Music' are published by arrangement with the *Daily Telegraph*. My thanks are also due to the British Broadcasting Corporation and the *Ampleforth Journal* for permission to publish the articles attributed to them.

I should also like to express my utmost gratitude to John Warrack for his generosity in giving me both help and advice throughout the whole preparation of this book.

D.C.

CONTENTS

PART I   PERSONAL

PART II   COMPOSERS

PART III   THE WORLD OF MUSIC: ARTICLES FROM THE *DAILY TELEGRAPH*

PART IV   SONGS AND SINGERS

PART V   ON THE AIR:
CONTRIBUTIONS TO THE BBC

Contents xvii

PART VI POETS AND POETRY

PART VII RELIGIOUS THINKERS

I always loved this bare and lonely hill,
this quickset hedge that largely shuts from view
the far horizon. But as I sit and gaze,
limitless space beyond silences supreme
and deepest quiet throng my imagination;
so for a while my mind loses its fears.
And as the noisy wind rages among
these tree-tops, I compare its voice with that
infinity of silence; and I recall
eternity, past ages and the present,
living and clamorous. So my imagination
founders mid these immensities, and I
find shipwreck in such a sea as this a boon.

after Leopardi

PART I PERSONAL

1

THE CRITIC'S TASK*

A critic of the arts is by definition an amphibian—that is to say, a man who divides his life between two elements in which, ideally, he should be equally at home. He must know a great deal about the nature, the language, and the history of the art he is criticizing; and this knowledge must be to the greatest possible degree sympathetic, involving his whole personality and not simply his cognitive faculties. If possible, he should have practical experience of this art in some form—sing or play some instrument, if the art concerned is music—but artistic experience of this kind must necessarily be secondary to the art which is his profession, that of the writer. This is the art of expressing in words not simply his opinions of the works he is asked to criticize, but of handling the infinite variety of nuances implicit in words; in the choice and placing of an adjective; in the abruptness or gentleness of phrase with which a judgement is made; in the distribution of emphasis that enables the reader insensibly to grasp the quality and the salient characteristics of a work or a performance. For writing is also a form of composition, in which harmony and orchestration must be used to modify and add colour and variety to plain melodic statements.

Reduced to essentials, art criticism is the discussion of one art by the practitioner of another. As such, it has always been resented and critics have been commonly regarded as failed painters, sculptors, musicians, or whatever it may be, rather than as a race of amphibians uneasily poised between two arts and with something to offer to both. In the case of music criticism, there are special elements that add to the complexity of the situation. In the first place, there is the intangible, abstract nature of music itself, which makes it the least amenable of all

* BBC 21 September 1971.

the arts to verbal discussion. Then there is the additional problem that one large area of the field—that of vocal music—is intimately associated with the critic's own art of words; and finally there is the fact that music, like the drama, depends on performance: the composer's conception can only be realized with the help of other artists.

Given these three factors, it is hardly to be wondered at if the critics are tempted to concentrate their attention on the easier and more congenial elements which invite verbal discussion— namely the performance of music and the territory which it shares with literature—and to neglect the music itself. For to discuss music itself, there are only two forms of language available: the strictly technical, only intelligible to professionals and a minority of amateurs; and the metaphorical, whose vague, subjective character makes it an unsatisfactorily blunt instrument. The only satisfactory solution is to make a conflation of the two languages, the proportion of technical to metaphorical of course depending on the reader envisaged and on the personality, tastes, and background of the critic himself.

The critic who is writing either a book or an article for a musical publication, with unlimited time and virtually unlimited space at his command, has an entirely different task from that of the journalist-critic—daily, weekly, or monthly—whose time and space are severely limited and whose language must not presuppose more than a minimum of technical knowledge in the reader. Daily journalism, or 'overnight criticism', is by its very nature ephemeral and can in today's circumstances hardly ever be more than trivial—which is not to deny that, like other trivialities, it can be well or badly done. The most that the overnight critic can hope to do is to start fruitful ideas in his readers' minds: he can put up hares, but he cannot possibly hope to hunt them. The writer for a weekly or monthly non-specialist journal is faced with a choice. He can either devote himself to the chase of a single hare—develop, that is to say, a single idea that seems to him not too out of proportion to his space—or he can attempt the general assessment of a work or a performance. His choice will be determined by the nature of his interest in music.

If we discount the journalistic element, present in varying degrees in all music criticism except that contained in books or articles for learned journals or encyclopaedias, music critics may be divided according to the nature of their predominant interest in the art. (Clearly few writers are chemically pure specimens of a type so the types themselves are only convenient generalizations.)

The musicologist proper is a scholar who rarely devotes much time to occasional or ephemeral writing. Concerned with the actual language of music and its historical development, one could say that he is the grammarian or philologist of the art, less interested in judgements of value than in the constatation and arrangement of facts. Many musicologists today write of music as an enclosed, self-contained world obeying its own laws; and they prefer to deny or disregard the influence of personal, social, or economic factors which threaten the objective, scientific character of their work. Critics with musicological interests but no scientific ambitions take extra-musical factors into account and view all music against the social, political, and economic background of the age in which it was written, and the circumstances of the composer's life, character, and temperament. This attitude is intellectually less ambitious but more comprehensive and more humane, in that it treats music as a single facet of a civilization and music history as a branch of *Kulturgeschichte*; and it takes into account the greatest number of non-specialist readers, for whom music is neither a profession nor a scientific discipline but still an important—often overridingly important—means of emotional and intellectual gratification.

The critic whose chief interest lies in performance has an important part to play in the maintaining of standards and in keeping before the public the various qualities that distinguish the great artist from the efficient practitioner, real artistic worth from technical facility. On the other hand, critics who write from a narrowly technical knowledge of a single branch of music have generally proved to possess restricted powers of literary expression as well as restricted musical interests; and their doctrinaire habit of mind easily alienates the general reader. The interest that a critic shows in details of performance generally varies in accordance with his knowledge of the medium and the work concerned; and this accounts for the fact that pianists receive more detailed and more intelligent notice than string-players, and that singers are more often criticized for their interpretation (partly a matter of literary understanding and taste) than for their vocal technique.

A type of critic that has always existed but has recently come to occupy a larger place in the public eye is the prophet—by which I mean the writer dedicated not to the vision of musical history as a whole (and the individual composer or performer in particular) in a clear and steady light, but to the championship of a particular composer or school, or the propagation of a particular musical creed. It is natural that the present century, which has witnessed a major revolution in the arts, should have produced more of these prophet- or champion-critics than any other; and many of them have done a great service to music by forcing orchestras and concert-giving institutions to take notice of works or composers that have been unjustly neglected. The gradual integration into the concert repertory of twentieth-century music which had found it difficult to obtain a hearing is largely the work of such champions. Whether or not the public has come—or eventually will come—to share their enthusiasm for the music which they have championed is less important than the fact that the conspiracy of silence—a conspiracy of laziness, prejudice, and ignorance—has been broken, and the musical public given an opportunity to form its own judgement.

The early stages of the twentieth-century musical revolution provided critics with almost unexampled opportunities to exercise their powers of analysis and exegesis, and their gifts as prophets and preachers. New developments of the language and a new aesthetic needed explaining and placing in perspective. Now, however, the situation has changed and one of the chief reasons for the critic's very existence is becoming questionable. Any criticism that can claim to be something more than the expression of a personal reaction or preference presupposes commonly accepted criteria; and such criteria in their turn presuppose the existence of a generally accepted language. In many works today this commonly accepted ground has been reduced to the instruments used: the uses to which they are put and the nature of the total aesthetic experience represents a deliberate break with hitherto accepted practice. In such cases the critic is thrown back on his powers of description i.e. what in fact happened—and a vague suggestion of whether he thought it 'a good thing' or not. This is hardly a profession; and in fact the whole movement among many of today's composers, notably Stockhausen, to reintegrate music into everyday life and to give concerts the airs of a political meeting, a sociological demonstration, or a religious service, removes them from the sphere of music criticism. Anti-art would suggest anti-criticism; and 'happenings' or 'events' involving music seem to invite the attention of a musically inclined reporter rather than criticism of any kind.

This is not a complaint, but simply the constatation of what appears to me to be a fact. Meanwhile there is plenty left for the critic to do. Anti-art and 'sociological' music represent only a minute proportion of the music performed in public; and the repertory is being steadily expanded, not only by new works which still fall very much within the music critic's field, but also by rediscoveries and revivals from the past. If there is a growing danger of music becoming a museum art, it is not a new danger.

Much more serious in my opinion is the fact that, in today's climate of subjectivism, judgements of value are becoming increasingly unpopular, even in the classical repertory. Critics are more and more regarded in some quarters as PR men, valued for their assistance in promoting music but resented if they express opinions that are unfavourable or, in the narrow sense, 'critical'. Not long ago I found myself taking part in a television discussion where I was in a minority of one in expressing a negative opinion of a work by no means new. A critic should be used to this, and it did not worry me unduly, although I was disturbed by the fact that two members of the group had privately expressed their agreement with me. What did concern me was the fact that one member turned to me and asked incredulously: 'But don't you realize that this work is a great popular favourite now?'—for all the world as though it were my obvious duty as a critic to echo the public rather than express my own considered opinion. It is this pressure on the critic to become a cheer-leader or a PR man—rather than the experiments of the avant-garde—that represents the chief threat to the continued usefulness and, in the last resort, the very existence of music criticism today.

2

EGON WELLESZ AS MUSICIAN AND TEACHER*

When Egon Wellesz died, on 9 November 1974, he had just entered his ninetieth year, a life-span unusual even today and one that made him the witness of a particularly rich and, in the literal sense of the word, catastrophic period of Western European culture. Brahms had still another twelve years to live in the Vienna where Wellesz was born; Liszt was still living; and Verdi had not yet written *Otello* or *Falstaff* nor Tchaikovsky his two last symphonies. Wellesz was to play a part in the last flowering of that specifically Central European culture which marked the last two decades of the old Austro-Hungarian monarchy and coincided with his own young manhood; to live through the humiliation and gradual reintegration of Austria, shorn of her political power but retaining much of her cultural significance; and finally, in his fifty-fifth year, to embark on an exiled existence in this country, where he found not only a welcome but honour, and the possibility of devoting himself wholeheartedly to the things which were closest to his heart—composition and teaching. He can hardly have imagined in 1939, when he was offered a fellowship at Lincoln College, Oxford that he was soon to enter on a period of his life when—in his sixties, his seventies, and even his eighties—his creative powers were to enjoy an almost unexampled 'second spring'.

His gifts were as various as his sympathies. Before he was thirty, he had already made sufficient mark with his studies of the baroque opera in Vienna to be appointed lecturer in musical history at Vienna University; and this interest in musicology was to persist throughout his life as parallel and complementary to his activity as a composer. In both fields he had the good fortune to find teachers of the highest quality—Guido Adler initiated

* BBC 17 November 1975.

him in the principles and practice of musical research; and between 1904 and 1906 his studies in counterpoint, fugue, and composition were directed by Arnold Schoenberg. And Schoenberg was to remain a unique influence in his musical development long after Wellesz, on Bruno Walter's advice, had embarked on his own, less revolutionary and more consciously history-orientated course as a composer. Perhaps a stronger, ultimately more formative figure in Wellesz's development than either Adler or Schoenberg was Gustav Mahler, whose career at the Vienna Opera coincided with the most impressionable years of Wellesz's early manhood. It was from the operatic performances given under Mahler's direction that he drew inspiration not only for his own operas but for that vision of musical history as a continuously evolving organism, that vision which he was eventually to impart to his pupils.

In one field especially Wellesz initiated invaluable original research and that was Byzantine church music, still hardly more than sporadically explored until he and H. J. W. Tillyard reached, independently, similar conclusions on the all-important question of deciphering the Byzantine notation. Apart from articles in learned reviews, published in some cases before 1920, the results of Wellesz's detailed studies appeared in two major volumes published in 1947 and 1948—*Eastern Elements in Western Chant* and *A History of Byzantine Music and Hymnography*. Characteristically, this scholarly interest stimulated and partly determined the nature of many of his early compositions. Of the five operas and four ballets which Wellesz wrote during the 1920s, two operas and one ballet are on ancient Greek themes: the opera *Alkestis* and the ballet *Achilles auf Skyros* have, respectively, libretto and scenario by Hugo von Hofmannsthal, Strauss's librettist and Wellesz's personal friend, while he himself adapted the libretto for his *Die Bakchantinnen* from Euripides's *Bacchae*. He also drew on Indian and Aztec legends, and the implicitly religious nature of these works for the theatre is not difficult to relate to his realization of the ancient Byzantine conception of the liturgy. Any relationship with the great spectacles of baroque opera at the Viennese court in the seventeenth century however is hard to imagine.

The musical style of these works reflects the same determining forces—the language is fundamentally tonal, like a slightly extended Strauss; dance rhythms play an important part; and it is only in the vocal line that the continuing influence of Schoenberg is clear, though it is modified by that of baroque coloratura. In the liturgical motets, mass-settings, and cantata which followed in the 1930s, Wellesz's style of composition took on an increasingly, and consciously, Austrian character which was to be further developed in the years of his exile in this country.

Meanwhile, however, perhaps owing to the steadily deteriorating political situation, composition played a lesser role among Wellesz's activities; and teaching, which was to form an important part of his life in Oxford, began to take its place. I am not certain who was his first English pupil in Vienna, but I think it was Spike Hughes—who made his name first as a kind of linkman between the worlds of classical music and jazz and has since contributed much to the study and popularization of opera in general and the history of Glyndebourne in particular. He was followed by two composers—Grace Williams and Dorothy Gow. I am not certain whether H. C. Colles, the chief music critic of *The Times*, was in their cases the agent; but it was certainly Colles who sent me to Wellesz in the last weeks of 1931, with the idea of my possibly joining his staff on *The Times* when I returned. In that prospect I was to be disappointed, but I have been eternally grateful to Colles for giving me the opportunity of spending two years in Vienna in constant touch with the kind of generally diffused but very precisely documented culture that radiated from both Egon Wellesz and from his wife, a distinguished art historian and a pupil of the great Sztrygowski.

Musically I was, at twenty-two, raw and almost ignorant except for the piano repertory with which I had a superficial acquaintance. Brought up in Yorkshire, I had had little opportunity to hear symphonic music of any kind and my only experience of chamber music had been at the Oxford University Music Club. Opera had meant occasional Carl Rosa performances and a handful of visits to Covent Garden. I suspect that it was my training as a classical scholar and a certain facility in modern languages rather than my musical enthusiasm that persuaded Wellesz to accept me as a pupil. In fact, when he discovered that I had studied Greek to scholarship standard, he tried for a short time to persuade me to specialize there and then in his own Byzantine field. Disappointed in this, he cheerfully and generously embarked on my general musical education. Although he was later to become a fluent English-speaker, he then knew none and my knowledge of German was hardly more than that needed for polite general conversation. How much I understood during the first months of the two years I spent with him, I don't know; but I was immediately stimulated—and as it were tantalized—by an awareness of the great fund of encyclopaedic knowledge and first-hand experience of cultural history with which I found myself in touch. Certainly I must have deceived Wellesz into thinking that I understood more than in fact was the case for he never spoke down to me in any way. He also took a great deal for granted in a twenty-two-year-old English boy. I remember him drawing some parallel between music and poetry and referring to the difference between the *weiche Fügung* in Rilke's poetic language and the *harte Fügung* in that of Stefan George. Well, I memorized the expression easily enough and was already familiar with Rilke; but my dictionary was not helpful nor were the other inmates of the Pension Keller, where Wellesz had arranged for me to live.

I have often wondered how much he can have known of the life and habits of the guests at what was nicknamed the 'Pension Atonal'. Certainly the cooking was excellent, good enough in fact to draw Alban Berg to lunch there with his wife two or three times a week, and his sister Smaragda even more often. For the owner of the pension was the May Keller whose relationship with Smaragda Berg had caused such gossip and even formed an obstacle to Berg's marriage; and the pension was in fact a kind of ante-room to the waiting-room of the psychoanalyst Dr Stekel. It was inhabited for the most part by a swiftly changing clientele of women of all ages and nationalities but—as even I learned before long—united in their homosexual proclivities. There were some interesting and distinguished women among them, but despite the excellence of the food the pension soon became too expensive for me and I moved to humbler and more workaday surroundings.

The plan of studies mapped out for me by Wellesz consisted of score-reading my way through the history of Western European music by day and nightly attendance at either an opera or a concert. At my weekly lesson we discussed what I had read and heard, each of my questions or answers giving rise to an impromptu lecture which often proliferated in the most unexpected directions. The name which recurred most often perhaps was Mahler's—for Mahler the conductor had been the hero of his youth just as Mahler the composer was to provide the starting-point for so much of the symphonic music which he wrote after 1943.

Although Wellesz's opera *Die Bakchantinnen* was given occasionally during the two years I spent in Vienna, he never spoke to me about his own compositions and I never thought of him as a composer. Although he had parted company with the serial Schoenberg, he was very anxious that I should attend the series of lectures that Webern gave one winter; and I duly went to these catacomb-like meetings held, if I remember right, above a chemist's shop and attended by hardly more than a score of people.

Looking back much later on the course of study Wellesz set me, I was struck by the fact that it did not include any examination of Schoenberg's music, virtually never performed in Vienna at that time. Events—in the shape of Wellesz's own compositions—were to make it clear, I think, that it was the Expressionistic aesthetic of Schoenberg's music rather than its revolutionary structure and language that concerned Wellesz as a composer. And a remark which he made to me towards the end of his life casts a retrospective light on this attitude. 'Yes,' he said, 'these young composers have learned to copy Schoenberg's manner—but the trouble is that they are no Schoenbergs'—as though he felt, as his own music implied, that to imitate the manner of so personal and unusual an artist was something to be avoided; and that to learn from him did not imply such imitation. He was in fact saying of Schoenberg what he had said of Wagner in one of his lectures on opera given in this country during 1933. After comparing Wagner's influence on music to that of Michelangelo on painting, he went on: 'In the same way, it has taken fifty years to overcome the pressure of his (Wagner's) personality on style and taste, to open up new ways of seeing and hearing, and to make it possible for us to stand free and untrammelled before our problems, enriched but not bound by the experience of the intervening years!'

And so in the nine symphonies that represent the major part

of his output in England and which were written between the ages of sixty and eighty-five, we find Wellesz working in the Austrian musical tradition whose origins in the Baroque age had formed the object of his earliest musicological studies. The influence of Mahler's music grew more marked as he grew older, but each of these works show also—in different degrees— close affinities with what Wilfrid Mellers has called the 'sacral nobility' of Bruckner and, most often in the scherzo movements, with the nervous, jagged lines of Schoenberg.

It is a natural tendency in an artist who finds himself an unwilling exile from his own country to become more conscious of his true artistic origins; and this tendency is reinforced by advancing age, when the awareness of the distant past often becomes sharper while more recent layers of experience lose vividness. In the years since the war, the passionate desire to rescue and revivify elements of the vanished Austrian past— memories of memories, in some cases, and all the stronger for that—became increasingly clear in Wellesz's music; and they earned him the gratitude and recognition, official as well as unofficial, of his fellow-countrymen.

In spite of gallant championship, particularly by the City of Birmingham Symphony Orchestra, none of Wellesz's music made any lasting mark in this country; and this not unnaturally puzzled and grieved him. I think that the reasons are not far to seek. In the first place, the public in this country has had little or no chance of hearing either Wellesz's early operas (the single opera that he wrote here, *Incognita*, was not representative) or the liturgical choral music written for the Catholic rite. Lacking the historical awareness which may perhaps be presumed in Austrian music-lovers and thus unaware of the links between Wellesz and his Austrian predecessors, the ordinary British music-lover found Wellesz's instrumental music lacking in the single quality which he most loves—strong and individual personality. It is this which has characterized all music, whether native or not, that has won recognition here. A strong personal flavour is almost the only characteristic shared by the composers who, at different times and in different degrees, have achieved popularity during the last half-century—Delius, Vaughan Williams, and Britten among the English composers; and Sibelius, Mahler, and Berlioz from abroad. It is not that the

composer of a symphony avowedly written 'contra torrentem'—
against the flood—and of the unaccompanied choral *Laus
Nocturna* lacked spirit and vigour. It was rather that, like his
older French contemporary and fellow-scholar Maurice
Emmanuel, Wellesz presupposed in his music an audience of
cultivated connoisseurs, in addressing whom it was not necessary
to raise his voice or to labour a point. Careful craftsmanship and
seriously reasoned thinking, a half-smiling reference to a shared
musical past, and a refined sensibility mark his chamber music
and much of his writing for solo voice—*The Leaden and the
Golden Echo* is a good example. His largest gestures are,
naturally enough, to be found in his operas and in some of his
church music; but even they take for granted a listener at home in
the atmosphere of classical tragedy (he liked to compare himself
to Gluck) or in the tradition of Christian devotion, neither very
common among audiences in the second half of the twentieth
century. And so in England, the memory of him as a composer
may be cherished by a minority; but all music-lovers will
recognize their debt to him as a scholar and a teacher, one who
did much to leaven the nervous insularity of the English musical
establishment of the day and to prepare and assist this country's
entry into musical Europe.

PART II COMPOSERS

3

GIACOMO MEYERBEER[*]

A hundred years after Meyerbeer's death his name can be read a hundred times in the pages of opera-histories for every single time that it appears on the hoardings outside an opera-house. It is, indeed, a name generally dishonoured among musicians, though more by vague generalization than by detailed criticism. The aesthetic canons of today are as different as possible from those of the French 'Grand Opera', which Meyerbeer stabilized and perfected, if he did not invent; and although those canons in many ways resemble those of late seventeenth-century Italian opera—which laid quite as much emphasis on the spectacular element—the French 1830s have decidedly less historical interest for scholars. And so historians are usually content to saddle Meyerbeer with the responsibility for the most conventional passages in the early or middle works of Wagner, Verdi, and Bizet without a thought of how this convention was first formed and then came to be accepted by men of such intrinsic originality. Every age numbers some Meyerbeers among its most admired composers—artists, I mean, who will be remembered less for their own works than for the influence they exercised on their successors, for having created the convention which provided the starting-point of new discoveries.

Meyerbeer was a cosmopolitan, and this accounted for much of his music's popularity during his lifetime and for its disappearance so soon after his death. Like Handel, he was German by birth and Italian by early adoption and training, yet eventually exercised his mature powers in still another country, whose traditions and tastes played a large part in determining his mature style. By 1864, when he died, national self-consciousness was already strong in the music of Wagner, in the

[*] BBC 9 July 1964.

then Bohemia, and in Russia; and six years later the Franco-Prussian war stirred a wave of this same nationalist feeling even in France, where Meyerbeer's music had been most admired and performed. Meyerbeer disappeared from pre-eminence with the Second Empire, and Western Europe was not to acclaim another cosmopolitan composer until Igor Stravinsky's Russian-French-American career came to repeat the pattern laid down by Handel and by Meyerbeer.

The world into which Meyerbeer was born—on 5 September 1791—was that which also produced in the same generation the poet Heinrich Heine and, less than twenty years later, Felix Mendelssohn. Meyerbeer's father, Jakob Herz Beer—a member of a Jewish banking family in Frankfurt—had established himself in Berlin, where he married Amalie Wulf, daughter of another successful Jewish business house, with many international connections and a strong interest in the arts. As the eldest of four sons in a rich and cultivated family, Jakob, who later added 'Meyer' to his name at the request of an uncle, was given every encouragement, including the best teachers available, when he showed his remarkable musical gifts. He studied the piano with Franz Lauska and composition with Carl Friedrich Zelter, the friend of Goethe who was also to be Mendelssohn's teacher, and with Bernhard Anselm Weber. This Weber—no relation of Carl Maria—was a friend of the famous or notorious Abbé Vogler with whom he toured as a pianist, and he had profited during a stay in Vienna from the friendship and advice of Salieri, the rival of Mozart and the friend of Haydn and Beethoven. Before Meyerbeer was twenty, therefore, he was in contact, direct or indirect, with most of the distinguished musicians in the northern and central European musical world. When in 1810, on his master Weber's advice, he was sent to Darmstadt to study with Vogler himself, he found as one of his fellow-pupils his contemporary Carl Maria von Weber, who was to play so important a part in the eventual creation of German opera. Meyerbeer, already a highly gifted pianist, here continued his studies in composition which issued, after two years, in some settings of Klopstock, and an oratorio entitled *Gott und die Natur*, which was performed at Darmstadt and obtained him the position of court-composer to the Duke. But he was already attracted to the opera, and it was the failure of this first attempt in this field—the biblical *Jephthas Gelübde* given at Munich in 1812—and the comparative success of his comedy, *Wirt und Gast* at Stuttgart the following year that proved turning-points in his career. The music of *Jephthas Gelübde* is blameless and dull, but *Wirt und Gast*—aided no doubt by Meyerbeer's excellent connections and long purse—was accepted for production in 1813 at the Kärntnerthor Theatre in Vienna where Meyerbeer went to oversee its preparation and also to make an appearance as a pianist.

The account in his letters of his journey down the Danube by steamer, delayed only by the bad weather and a truly extraordinary number of amorous adventures, shows Meyerbeer at twenty-two gifted, handsome, rich, ambitious, and lacking only a sense of direction—the lack that was, in a sense, to be his undoing all along. Meanwhile, however, a concert of Hummel's which he attended the evening he arrived in Vienna, made such an impression on him that he set aside the first months of his stay in Vienna to re-fashioning his own technique. His recital, when it took place, won him golden opinions, among others those of Moscheles, and there is little doubt that Meyerbeer could have made a career as a concert pianist. His opera, on the other hand, was a failure; and he seems to have felt a musical provincial in Vienna—a general sense of humiliation that was objectified on the famous occasion (which we have no reason for believing apocryphal) when he played the bass drum in a performance of Beethoven's *Battle of Vittoria* and was pulled up by the composer with the harsh but shrewd comment that 'he never had the guts to come in properly at the right time'.

It may well have been to escape this feeling of failure that on leaving Vienna, Meyerbeer paid his first visit to Paris. Very little is known of these months spent in Paris during 1815, but there can be no doubt that his impressions of the Bourbon Restoration and of the Spontini regime at the Opéra further stimulated his ambition. Nevertheless he was still, beneath his sophistication, a puzzled and undecided youth and something of a chameleon; and Salieri's advice to him in Vienna—to go to Italy and study opera in its home country—easily persuaded him. Italian music in the spring of 1816 meant Rossini; and forty years later Meyerbeer was to remember the enchantments of those first months:

I was attracted—he wrote to Dr Schucht—quite independently of my will, by these delicate meshes of sound [Rossini's music, that is]. I seemed to be imprisoned in a magic park from which I neither could nor would escape. All my faculties, all my thoughts were becoming Italian; after I had lived there a year I felt like an Italian born. . . . That so complete a transformation of my inner life should have the most essential influence on my style of composition may be readily understood. I did not wish, as people imagine, to imitate Rossini or to write in the Italian manner, but I was obliged to compose in the style that I adopted *because my state of mind compelled me to do so.*

Meyerbeer's 'state of mind' was, indeed, a stronger factor than either his musical character or his musical convictions—both of which were weak—and between 1817 and 1824 he wrote six operas indistinguishable from those of any Rossini-imitator and correspondingly popular in Padua, Turin, Venice, and Milan where they were given. His old friend Carl Maria von Weber, who had confided in Meyerbeer his hopes for a national German opera, was heart-broken and wrote frankly: 'It makes my heart bleed to see a composer of creative ability stoop to become an imitator in order to win the favour of the crowd.' The accusation coming so early in Meyerbeer's career and from so sincere a friend, is significant.

Before the last, and certainly the most interesting, of these Italian operas—*Il Crociato in Egitto*, given in Venice in 1824—Meyerbeer returned for a short time to Berlin; but his ambition, only temporarily satisfied by what were after all the merely provincial successes that he had enjoyed in Italy, prompted him to return to Paris; and in the event it was Rossini himself who made this possible, by offering to stage *Il Crociato in Egitto* at the Théâtre Louvois. Without the castrato Velluti, the last of his kind, for whom Meyerbeer had written a principal part, *Il Crociato in Egitto* was not a success in Paris; but Meyerbeer's time there was crucial because during it he met the librettist Eugène Scribe, who had already started his immensely successful collaboration with Auber.

When Meyerbeer returned in 1826 to Berlin, to his father's death-bed, he took with him the first draft of what was, five years later, to be *Robert le Diable*. This was announced as early as May 1827 as 'a comic opera in three acts'. But the next years were to be filled for Meyerbeer with too many personal events for him to devote much time to composition. Soon after his father's death he married his cousin, Minna Mosson; and the death of his boyhood friend Carl Maria von Weber in 1826 was followed in tragically quick succession by the deaths of the first two children that Minna bore. And so it was not until 1830 that Meyerbeer returned to Paris with the completed score of *Robert le Diable*. Its performance was postponed first by the July Revolution and secondly by the composer's unexampled demands in the way of rehearsals, which lasted for five months. For Meyerbeer, at nearly forty, had had enough of qualified successes and was determined to leave no detail to chance, no effect unconsidered either in the music or the production. It may be that his domestic tragedy—so like that which Verdi suffered in his first marriage—had broken Meyerbeer's nerve; but there is no doubt that, for whatever reason, he had already developed many traits of that anxiety neurosis which was to haunt him for the rest of his life and to embitter even his greatest successes. Pathologically shy and nervous, he spent such money as did not go on subsidizing production of his works, in the purchase of privacy, eventually developing such a horror of being buried alive that he left instructions that he was not to be buried for several days after his death and that bells were to be attached to his hands and his feet. His pursuit of popular success may be regarded as another facet of this desire for reassurance.

If I have so far dealt in some detail with Meyerbeer's life, it is in order to show what were the musical influences to which he was exposed as a young and impressionable man, and to search for any indications of how we should explain the character of the four big operas which engaged him for the remaining thirty years of his life and form the basis of his reputation. Although solidly grounded by Zelter and Bernhard Anselm Weber, he had turned instinctively to the more liberal and adventurous harmonic teaching of the Abbé Vogler. This permitted a then unexampled freedom in the use of chords of the seventh, ninth, and eleventh on any degree of the scale, without modulation, and even countenanced chromatic alterations of these chords and their inversions. There was more than a streak of the charlatan in the old Abbé, who indulged in mystifications and exotic experiments of doubtful artistic character in his organ recitals—thunder storms, the *Fall of Jericho* or *Variations on a Hottentot theme*—and even in his compositions. The fact that these mild charlatanries impressed the public and increased Vogler's reputation was certainly not lost on so bright-witted and ambitious a boy as Meyerbeer.

His sense of failure when he first competed in the international musical arena at Vienna and his easy successes in Italy, as little more than a pasticheur of Rossini, had combined to make him cynical about the public, whose approval he nevertheless craved, with all that instinctive, illogical craving that can afflict a natural 'insider' who by the mere accident of birth finds himself marked from the start as an 'outsider'. This was in fact the position of every member of the Jewish community, and it must be borne in mind, whether we are seeking to explain Mendelssohn's wonderfully successful, apparently instinctive conformism, Heine's exhibitionist aggressiveness, or Meyerbeer's anxious but indomitable determination to bribe his way out of the golden ghetto into which he was born.

It happened that events in Paris provided him with just the opportunity that he needed, at exactly the right moment in his career. The last superintendent of the Parisian theatres before the July Revolution was a certain Vicomte Sosthène de la Rochefoucauld, immortalized by Théophile Gautier as 'the unfortunate and virginal viscount who lengthened the skirts of the dancers at the Opéra and with his own patrician hands applied a modest plaster to the middle regions of all the statues'. His regime was not unnaturally a financial disaster for the Opéra, despite two enormously successful productions, Auber's *La Muette de Portici* (or *Masaniello*) in 1828 and Rossini's *Guillaume Tell* in 1829. The first artistic concern of the new regime of Louis-Philippe was to ensure the Opéra's solvency; and its administration was therefore, for the first time, entrusted to a 'director-entrepreneur, who shall manage it for six years at his own risk and fortune'. It was in fact farmed out for private exploitation; and the new director was frank in his aims. Dr Louis Véron had considerable experience in the fields of publicity, including journalism and medicine, and he tells us in his memoirs[1] that it was this experience, rather than any (in fact

---

[1] Véron, Louis, *Mémoires d'un bourgeois de Paris*, 1856–7.

non-existent) musical qualifications, that he urged at his interview with the Minister of the Interior:

I was permitted to explain in a few words how a brilliant and skilful direction of the Opéra might be valuable politically at the beginning of a new reign. It was desirable that the foreigner should be attracted to Paris by the fine performance of musical masterpieces, and that he should find the boxes filled with an elegant and tranquil society. The success and receipts of the Opéra should be a testimony to the government's stability.

Véron's candidature was backed by an advance of 200,000 francs from Rossini's close friend, the Spanish banker Aguado, and once successful, Véron embarked on a publicity campaign worthy of the great American art-tycoons whom he anticipated in so many other ways. His whole existence became advertising copy—'his carriage, his horses, his dinners, even his cravats' and of course his personal relationships of all kinds. He surrounded himself with a court of influential businessmen and literary hangers-on, among whom he paid especial attention to the journalists on whom he relied for publicity and, in the event, for favourable notices. The kind of works that he was to give to the public depended, of course, on what public he wished to attract. The social upheaval following the July Revolution had brought prosperity to a new and powerful middle class—'men of business, men in the professions, politicians seeking distraction from their daily worries and anxious to prove that money could take the place of birth'. These were in no sense connoisseurs of music, but neither were they complete Philistines. They were, however, enormously open to the appeal of what would now be called 'glamour'—the suggestion of a world of fabulous sensual delights and easy emotional excitements, of physical splendour and aristocratic magnificence all within their reach for the expenditure of a comparatively small fraction of their new-found money. Véron skilfully sharpened public interest by removing the 'no admittance' sign from the stage door of the Opéra and making the Opéra balls as luxurious and as easy to attend as possible. This was the world inhabited and described by Balzac,[2] who must indeed have known it well, since he was one of the *habitués* of the famous *loge infernale* who were said to have had opera-glasses made 'that enlarged objects thirty-two times and from which the dancers' tights kept no secrets'.

[2] See *Gambara* (1839) and *Massimilla Doni* (1839).

The opera for such a public must clearly be spectacular in every way. Only lavish display of vocal and every other kind of charm would convince the aesthetically simple-minded, purse-proud bourgeois that he was getting his money's worth; and Véron already had pointers to the direction in which public taste could be exploited. Rossini had been called in to regenerate the art of singing in France as early as 1824, when he became director of the Théâtre Italien; and with the example of the great Italian singers whom he introduced to Paris, a new generation of French artists was growing up—including the sopranos Cinti-Damoreau, Maria Malibran, and Dorus-Gras, the tenor Adolphe Nourrit, and the bass Nicolas Levasseur. It was singers of this calibre, well qualified to deal with Rossini's spectacularly florid, yet strictly defined vocal writing, that made the success of his *Mosè in Egitto*, *Le Comte Ory*, and *Guillaume Tell* and of that most prophetic of all French operas of the 1820s—Auber's *La Muette de Portici*. The revolution in the scenery and production at the Opéra, demanded by the tastes of the new public, had already been brought about by Louis Daguerre (inventor of the daguerrotype) who specialized in panoramic lighting-effects; Edmond Duponchel, an enthusiastic and ingenious inventor of magnificent historical sets; and Pierre Cicéri, who had even been sent by the government to a performance of Pacini's *Last Day of Pompeii* to observe La Scala's presentation of the eruption of Vesuvius, which appears also in Auber's *La Muette de Portici*. Véron's chief conductor at the Opéra was the same Habeneck who took such pains with the first performances of Beethoven's symphonies in Paris, while the ballet was led by Marie Taglioni and Fanny Elssler. Véron, characteristically, economized by lowering the pay of the rank and file of the Opéra—members of the orchestra, the chorus, and the *corps de ballet*—but never objected to the high fees demanded by his principals.

Since Meyerbeer left Paris not long after the failure of his *Il Crociato in Egitto* in September 1825 and did not return until 1830, he was not in fact present at the first performances of either *La Muette de Portici* or *Guillaume Tell*, though he undoubtedly saw them later and knew their music. In any case he had in Eugène Scribe a friend and a collaborator who was intimately connected with the new regime at the Opéra and librettist of *La Muette de Portici*. Meyerbeer and Scribe agreed in rejecting both classical myths and already existing stage plays as operatic subjects. Scribe himself was an enormously adroit theatrical carpenter, with the eye of a lynx for the vagaries of public taste, and a psychological flair for what the new theatrical public would enjoy and understand. A taste for the Gothic tale of 'mystery and imagination' which had invaded France well before the Revolution of 1789 received a new stimulus during the Restoration, when the public no longer had the exciting realities of the Napoleonic wars to occupy their imagination. The popularity of Weber's *Der Freischütz*, even in its mangled form as *Robin des Bois*, and Boieldieu's *La Dame Blanche*, with a libretto by Scribe based on Walter Scott's *The Monastery*, are clear evidence of this taste in the opera-house. That Rossini could enjoy such a success with *Le Comte Ory*, in which Scribe parodies the whole Restoration taste for medieval, Walter Scottish pageantry, and sentimentality, is a sure indication of how strong this taste had been. The grotesque and macabre tales of E. T. A. Hoffmann, too, which Meyerbeer doubtless knew in the original German, obtained an instantaneous success in France when they were translated. In fact the opera with which Meyerbeer returned to Paris from Berlin and which had its first performance on 21 November 1831, after five gruelling months of rehearsal, during which Meyerbeer took advice from any member of the Opéra staff who chose to proffer it, was an extremely clever concoction designed to appeal to all these shifting currents of taste and to provide something novel and piquant into the bargain.

When Wagner later attacked Meyerbeer in his notorious pamphlet, *Das Judentum in der Musik*, his chief charge was that, from sheer Jewish rootlessness and lack of national tradition, Meyerbeer had for the first time made 'what the public wanted' the primary consideration of his operatic aesthetic; that whereas many other composers had borne popular taste in mind and allowed it to modify their musical utterances, in Meyerbeer's case this 'popularity' was the sole determining factor. It would of course be quite possible to admit this to be true of Meyerbeer—and the enormous popularity of his works with one generation, and their swift disappearance from favour when taste changed, would bear out Wagner's theory at least in part—without accepting Wagner's ridiculous generalizations about composers of Jewish origins. It is certainly quite as ridiculous to speak as though the peculiar position in which sons of rich Jewish banking-houses in Germany—such as Meyerbeer, Heine, and Mendelssohn—grew up during the first two decades of the nineteenth century had no effect at all on their characters. A ghetto is none the less a ghetto for being paved with gold and lined with bank-notes; and to find every door open to one's money except one—and that the door leading to social acceptance as an equal—is a bitter and humiliating experience that will leave no young man of spirit unmarked. Heine's whole life was a tortured shuffling between defiance and unsuccessful attempts to take on French national colour as a protection. And he showed his worst side to his own people such as Meyerbeer, from whom he accepted much-needed money to act as a kind of publicity agent, only to turn and make fun of his patron and his pretensions. Mendelssohn's apparently instinctive and flawless conformism was achieved, perhaps, by refusing to penetrate below the surface of a happy family life and a successful career; or alternatively his character and his genius may have been so well adjusted to each other, and to the purely formal demands of his art, that the impulse to confront the shadow-side of life—his own life or life in general—never appeared to him except as a symptom of illness or fatigue. Meyerbeer, on the other hand, when he came to write his *Robert le Diable* had nearly reached forty without the thirty years of 'promise'—as a young pianist, a grand ducal protégé, and a local Italian operatic light—ever having been crowned by unqualified success. He was the victim of tantalization, unsure (like many rich men) of how much he owed to his money even that measure of success that he had enjoyed; and spurred on, perhaps, by domestic tragedy to make sure once and for all of capturing the most important of all operatic publics.

When he saw the new conditions at the Paris Opéra, where money could buy everything—quality as well as quantity, lavish sets and productions as well as first-class singers and endless rehearsals, and good press notices into the bargain—he recognized his chance. I think it would be unrealistic to say that he was wholly uninfluenced by an instinct—an instinct more probably than a conscious desire—to conquer the society that accepted his money but in the last resort refused him, to compensate himself for the countless occasions when he had been made to feel different, an outsider. The operas that he wrote after 1830 are the works of an obsessional would-be insider, for they contain (as Mendelssohn said with distaste after seeing *Robert le Diable*) 'something for all tastes'—and that includes trained musicians and genuine music-lovers as well as all varieties of groundling. What Mendelssohn found these operas lacked was 'a heart'—by which he meant, no doubt in the first place, a 'true German heart'. And we must remember that he, like many Germans before and since, was humiliated and affronted by French cultural pretensions and by the style and glitter of Paris, which made him feel at the same time provincial and morally superior. Meyerbeer conquered Paris by wearing a French mask, just as Lully and Gluck had done before him. There was nothing disreputable in this, or in the Italian elements of his style, which could easily be paralleled in Gluck's Paris operas. His case was, however, different from Gluck's in two important ways—first in that he was writing for an uneducated and sensation-hungry public, whereas Gluck was writing for the court and the educated middle classes; and second, that Gluck, for all his flexibility and willingness to compromise on occasion, was a man of strong basic character and convictions, whereas Meyerbeer had neither. He commanded an excellent conventional technique of composition, which meant then, as now, a great facility in imitating accepted models. What was original in his music was not its actual character but the combination of ideas borrowed from Spontini, Rossini, and Auber and the theatrical effectiveness with which this combination was presented. For these operas are not primarily, still less exclusively, musical works; and it is unfair to Meyerbeer to judge them as such. They are in fact what Wagner called a *Gesamtkunstwerk*, though not quite in Wagner's sense—theatrical entertainments in which stage spectacle, ballet, lighting, sets, and 'machines' form quite as integral a part as singing and orchestral music. 'Musico-dramatic historical pageants' would be a clumsy but less misleading way of describing them than 'operas'; and the Baroque mythological musical entertainments given in seventeenth-century Italian courts provide the closest historical parallel.

The four operas on which Meyerbeer's reputation rests—or

perhaps I should say rested—can be considered together, from the stylistic point of view. They are *Robert le Diable*, produced in 1831; *Les Huguenots*, produced in 1836; *Le Prophète*, produced in 1849; and *L'Africaine*, produced after Meyerbeer's death in 1864, in the last of several versions, revised by Fétis (the only instance to my knowledge of a composer having his work 'finished', at least in this sense, by a critic). The other works written by Meyerbeer during this last period of his life need hardly detain us, but they should perhaps be mentioned here. One of them was connected with the official post of General-musikdirektor in Berlin, which Meyerbeer held from 1842–9. *Ein Feldlager in Schlesien*, written for Jenny Lind, deals with an imaginary incident in the life of Frederick the Great and is made the occasion for a glorification of the army. It was described by one critic as 'das typisch preussische Hoffestspiel, mit riesiger Aufmachung, militärischem Prunk, sentimentalem Hurrapatriotismus und familiärer Hohenzollernbeweihräucherung'—'a typical Prussian court work, with giant expense, military glitter, sentimental jingoism, and familiar adoration of the Hohenzollern dynasty'. Meyerbeer rewrote this work for Vienna as *Vielka* and for Paris as *L'Etoile du Nord*, in which form it was admired by Berlioz in 1854. He found it 'wonderful in point of truth, elegance, and freshness of ideas. By the side of the most alluring, coquettish devices are to be found startling complications and striking touches of passionate expression.' The other minor work was also an *opéra comique*, *Dinorah* or *Le Pardon de Ploërmel*, remembered today if at all only for the 'shadow-song' beloved by an earlier generation of coloratura sopranos. *Dinorah*, however, takes us back to the first of the big four, *Robert le Diable*, where the influence of the old *opéra comique* is still strong.

Meyerbeer was well aware of the French middle class's attachment to their national form of opera, with its inheritance from the eighteenth century—wicked baron and virtuous milkmaid; peasant lover, also more or less virtuous but more noticeably a ninny or Sancho Panza figure; scenes from 'innocent' peasant life contrasted with the reprobate luxury of the rich; and a plentiful indulgence in primitive descriptive music—ding-dong choruses for the church bells, reapers' choruses in which they whet their scythes in the orchestra, and so forth. To combine this tradition with the new historical-tableau or pageant type of opera seemed an obvious recipe for success; and Meyerbeer used it in *Robert le Diable* and again, less obviously, in *Le Prophète*.

The characters of Alice and Raimbault in *Robert le Diable* are set quite apart from the grand world of Robert, Count of Normandy and his Mephistophelean protector Bertram. Alice is a Micaela, sent with a letter from his mother to reclaim Robert from his wicked ways; and Raimbault, who gets into trouble by singing a ballad about Robert the Devil, is made a butt for Bertram's grim humour—just as the cruel baron in the traditional *opéra comique* plays like a cat with the bumpkin whose betrothed he means to seduce (Don Giovanni and Masetto, in fact). In *Le Prophète* the situation of Berthe and Jean, the peasant lovers persecuted by the wicked Count Oberthal, is exactly that of Lucia and Renzo in Manzoni's *I Promessi Sposi*. Berthe and Jean's old mother, Fidès, stand out by their simple would-be unaffected rhythms and melodies, though Meyerbeer cannot resist giving these a showy and ingenious frame.

I wanted to mention this *opéra comique* element in Meyerbeer before discussing the general characteristics of his chief operas because it is found only in two of them and is very often disregarded by those who write about his music as though it were simply a highly seasoned ragout of Rossini and Auber. Each of the four chief operas is very long. Louis-Philippe's purse-proud bourgeoisie wanted a good run for their money, and plenty in quantity as well as what they regarded as quality. Scribe, who played a very large part in determining Meyerbeer's mature style, knew his public perfectly. They were to be given five acts—with a ballet, or chief scenic attraction in Act III and the sentimental *clou* in Act IV, which must culminate in one of the monster *morceaux d'ensemble*, with chorus added—such as the blessing of the daggers before the St Bartholomew massacre in *Les Huguenots* or the coronation of the prophet John in Munster Cathedral in *Le Prophète*. Act V brings the denoue-ment, with retribution for the guilty and death or quasi-apotheosis for the virtuous, and probably a transformation-scene, a spectacular and edifying death, or a *deus ex machina*, as in *Les Huguenots*.

Each scene of each act provides some kind of diversion, generally visual, for the groundlings—precisely what incited Mendelssohn to his scornful remark about 'something for all tastes'. *Robert le Diable* has an exciting gambling scene in the first act (not lost on Verdi or Massenet in *La Traviata* and *Manon*); a 'Valse infernale' and a titillating scene of mild sadistic character between Bertram and Alice, followed by the ballet of spectral dissolute nuns, who tempt the hero; a scene in which with the magic golden bough Robert first sends the Sicilian court to sleep and then wakes them; and finally a cathedral wedding. *Les Huguenots* and *Le Prophète* go much further in such matters. *Les Huguenots* opens with an 'orgy' of Catholic noblemen, enlivened by Marcel's *pifpaf* imitation of a battle. This is followed by a ballet of bathing beauties and a spectacular emotional scene in Act II; Protestant rataplan chorus and Catholic litanies, gypsy dances, curfew, duel, and wedding procession in Act III; blessing of daggers and love duet in Act IV; massacre and royal appearance in Act V. As the King of Prussia tartly observed of this opera 'Catholics and Protestants cut each other's throats and a Jew sets the proceedings to music'. Perhaps Meyerbeer's boldest inventions in this line were the ballet of skaters, the coronation scene, the solemn exorcism, and the catastrophic final explosion in *Le Prophète*. Selika's death under the upas-tree in *L'Africaine* is a tame conclusion to such a list.

Now there is no doubt at all that one of the main interests in these great jamborees is visual, for reasons that I have already gone into. But we cannot assume simply for this reason that the music is negligible. In fact it can hardly have been that, if it was admired so much by Berlioz, Bizet, Gounod, Verdi, Wagner, and Tchaikovsky. These men all had reservations about Meyerbeer's music, but they nevertheless paid it the sincerest of all compliments—the tribute of imitation, as I hope to show.

What in fact were Meyerbeer's gifts and qualifications as a composer? He was no great melodist—Vasco's 'O paradis sorti de l'onde' from *L'Africaine*, always quoted as evidence of his melodic gift, is the only solo number that has really survived from all four of these operas. For the most part Meyerbeer was content either to imitate Rossini and Auber (who was often himself imitating Rossini) or to borrow the melodic forms of the *opéra comique* and contrast their simplicity with the Italian-style decoration of the 'grand' airs. He was, however, skilful at concealing this poverty of melodic invention, either by striking instrumental accompaniment or by strongly stressed dramatic character. A good instance of the former is Raoul's opening air in the first act of *Les Huguenots*—'Plus blanche que la blanche hermine', which is accompanied by a single instrument and that the then unheard of viola d'amore. An instance of the strong dramatic character of melody masking its intrinsic poverty is Raimbault's ballade from *Robert le Diable* which Wagner plagiarized for Senta's ballad in *The Flying Dutchman*.

If Meyerbeer had no striking gift for inventing great solo melodies, such as opera has always thrived on, the same cannot be said of his ensembles and choruses, where he was supported by his remarkable harmonic sense and that ability to build up to a great dramatic climax from a single germ idea—a gift which has been compared, rather generously I think, to that of the symphonist. These great ensembles and choruses of Meyerbeer's are musically the most striking features of his operas—huge tableaux in which he may combine as many as seven principals and a large chorus, playing one body against the other like the ripieno and soloists in a concerto. One of the most striking of these is in the second act of *Les Huguenots*. The finale opens with representatives of Catholics and Protestants swearing eternal peace, at the instigation of Marguerite de Navarre, four male soloists with male chorus. The scene is introduced by pianissimo timpani (the parallel between this scene and the Prelude to *Rigoletto* cannot be fortuitous); otherwise the orchestra only plays with the chorus. The soloists sing in octave unison and then at the height of the excitement break into an unaccompanied prayer, upon which the chorus calls down heaven's blessing. This is followed immediately by Raoul's outrageous refusal of Valentine, which introduces the main body of the finale—always called by Meyerbeer 'strette', using the word in its secondary, general sense of 'bringing the accents closer together and thus producing a climax of excitement' as *Grove* puts it. Most magnificent of all these set-pieces, though not perhaps as strikingly unified in character as the one to which I have just referred, is the scene of the blessing of the daggers in Act IV. This is carried out by three monks, while the Catholic leaders stand by, and the note of fanaticism grows in the chorus as the monks move from group to group assuring them that the Catholic cause is just and that 'Dieu le veut!'—no heretic is to be spared. The dotted rhythm, and the recurring figure with which it is connected, are very characteristic of Meyerbeer, as are the theatrical alternations of the fortissimo chords of E major with pianissimo chords of A flat major—the sort of enharmonic sequence (it is not a modulation) that dazzled Meyerbeer's contemporaries. The thundering out of the E major theme in unison, with its threatening triplets was to become a commonplace (we find it as late as Verdi's *Otello*, in the duet which ends Act II), but must have made an overwhelming impression when it was first heard.

We have already referred on several occasions to Meyerbeer's harmony, and I want to say something more about this. He belongs of course—none more obviously—to what may be called the diminished-seventh school of operatic composers; but he went further than that. His harmonic boldness was, of course, strictly superficial and decorative, not in any sense organic, but it has been the rule of history that harmonic expansion has taken place by chords which were at first considered exotic or merely decorative being accepted eventually in their own right as organic; or by chords heard at first in what seemed to the ear bold succession eventually being accepted simultaneously. There is an interesting example of this in the exorcism scene in *Le Prophète*, where Jean, having publicly denied his own mother, forces her to admit that she was mistaken in claiming him as her child. First comes a fairly obvious repetition of the same phrase mounting by semitones (Puccini was to do the same thing on a much grander scale in *Turandot* of course); then, while Fidès is fighting with herself, a high tremolando in the strings introduces a sequence of chords whose roots lie an augmented fourth apart. It was to be just over fifty years before Ravel dared to place those chords in even closer juxtaposition in *Jeux d'eau* and another ten after that before Stravinsky positively relished their dissonance in *Petrushka*.

A great many of Meyerbeer's most effective harmonic *coups* are dramatically placed chords of what Gerald Abraham has called the 'Russian' sixth, owing to its popularity with Russian composers—the chord of the dominant seventh on the flattened sixth of the scale. There is a spectacular use of this chord in the cathedral scene in *Le Prophète* where it is characteristically employed to give interest to the melodically conventional solo of Fidès. The same harmony is prominent in 'O paradis' in *L'Africaine*. *Le Prophète* was first given in Paris in 1849, the year in which Liszt revised his A major Piano Concerto, which opens with exactly this progression. Whether the Russian composers were influenced by Meyerbeer (whose *Le Prophète* was first given there in 1852) or by Liszt's Concerto would be hard to say. The most famous instance of the progression, which became a commonplace with Rimsky-Korsakov and Tchaikovsky, is in the letter song from *Eugene Onegin*.

Another instance of this progression in a very different context throws a slightly sinister light on Meyerbeer's influence as a harmonist. Histories of music generally blame Gounod for introducing the particularly oleaginous kind of harmony connected with the more mawkish Victorian hymn-tunes; and in effect it was he who was responsible for introducing it into England in his much admired 'sacred' music—*La Rédemption*, *Mors et vita*, and a whole heap of 'sacred' songs. But where did he find it? There is no trace of chromatic harmony in his delicately written early works, *Philémon et Baucis* and *Sapho*. It is not until *Faust*, in 1859, that Meyerbeer's influence becomes strong in his music—in the whole character of Mephistopheles, in the church scene and the final 'apotheosis', as well as most unmistakably in the ballet. The Russian sixth harmony is crucial in Faust's 'Salut, demeure chaste et pure' and again in the Jewel Song. But what are we to say when we find Vasco in *L'Africaine* actually producing what is to all intents and purposes a full-scale Victorian hymn-tune of the Gounod type, only going off the ecclesiastical rails in the excursion into Russian-sixth territory. *L'Africaine* was produced in 1865 and it was in the 1870s, while he was living in London as a refugee from the Franco-Prussian war, that Gounod started his infiltration of cathedral organ-lofts and provincial festivals. The poison he brought with him—if we may speak melodramatically—was, I believe, of Meyerbeer's brewing. What it became in other hands we can see from Wagner's *Tristan*, whose harmony no doubt owes something to Spohr and to Chopin, but perhaps also something to Meyerbeer. One of the crucial harmonies of the Tristan love-duet is, indeed, just this same Russian sixth again, and in the same key and position as it occurs in the central scene of *Le Prophète*.

Only slightly less important than Meyerbeer's harmony is the influence of his orchestration. Spontini was the first to introduce the characteristically Napoleonic idea of orchestral grandiosity for its own sake. When *La Vestale* was produced in 1807, there was a joke current in Paris about a deaf man advised by his doctor to go to a performance of this opera to cure his malady. The two men went together to the Opéra and after a particularly loud orchestral outburst the deaf man turned in delight to his neighbour: 'Doctor', he cried, 'I can hear.' But the doctor did not answer; for what had cured his patient had deafened him. The Spontini tradition was followed by Auber in his *La Muette de Portici* and by Rossini in *Guillaume Tell* and was naturally accepted by Meyerbeer as part of his arsenal of 'effects' to astonish and delight his audience—or *épater les bourgeois*, as Théophile Gautier expressed it. The practice of marking each beat, or at least each strong beat in the bar, with the cymbals; the unleashing of sudden fortissimo chords of the dominant seventh in the brass; using the heavy brass to double the basses of the chorus, often with trombone triplets, or even with chords off the beat—all these practices of Meyerbeer's which persisted at least as late as the triumph scene in *Aïda* (1881) and Tchaikovsky's *Sleeping Beauty* (1889), where Carabosse's entry is pure Meyerbeer.

Meyerbeer had a special predilection for the bassoon, as Gluck had for the oboe, and Weber for the clarinet, and there are countless instances in these operas where the instrument's characteristic hollow, nasal tone and somehow unwieldy agility are spotlit. The most famous is no doubt the Anabaptist sermon at the beginning of *Le Prophète*—the 'Ad nos ad salutarem undam' that Liszt was to borrow for variations. The three preachers sing in octave unison with the bassoons.

The cor anglais solo in *Robert le Diable* and Meyerbeer's frequent use of solo timpani (as in the passage from *Les Huguenots* described above), bring us to the vexed question of the priority of novelties in orchestration between Berlioz and Meyerbeer. The two men were friends of a kind; Meyerbeer needed the goodwill of Berlioz the critic and Berlioz admired a great deal in *Les Huguenots* and *L'Etoile du Nord* as well as isolated features of other works. Probably his attitude is best expressed in the letter he wrote to his sister after the first performance of *Le Prophète* in 1849:

I hope that Meyerbeer has the good sense not to take amiss the four or five reservations I put into my ten columns of praise. I should have liked to spare him the pain . . . but there are certain things that must absolutely be said out loud . . . I cannot let anyone think that I approve, or even condone, the compromises that *such a master* [notice that expression] makes in favour of the bad taste of a part of the public . . . This score contains some very fine things side by side with feeble and detestable ones. But the splendour of the *show* will win acceptance for everything.

Meyerbeer probably became acquainted with the *Symphonie Fantastique* at the time that *Robert le Diable* was being rehearsed at the Opéra, that is to say, between June and November 1831. The conception of both works is so characteristic of the mood of the day that we do not need to suppose any further connection. Jacques Barzun makes the non-committal, but heavily weighted observation in a footnote that 'the opinion of those competent to judge, from César Cui to Arthur Hervey, supports the view that Meyerbeer drew heavily on Berlioz's melody, instrumentation, and dramatic conceptions'. Yet if we accept this, to say the least of it doubtful, proposition, we find ourselves attempting to explain the best of Meyerbeer in terms of Berlioz—a very difficult task—rather than the worst of Berlioz in terms of Meyerbeer, which is very much easier. Meyerbeer was certainly an inveterate borrower of ideas; but all his best scenes show exactly that purely theatrical skill that Berlioz never wholly achieved, or only after Meyerbeer's death, in *Les Troyens*. That Meyerbeer borrowed individual ideas of orchestration is probable—the cor anglais solo in *Robert le Diable* perhaps, the echoing 'pastoral' clarinets in the prelude to *Le Prophète* almost certainly.

That Meyerbeer, or anyone else, ever drew heavily on Berlioz's highly individual melodies seems to me out of the question. When we come to Berlioz's debt to Meyerbeer, we are confronted simply with a list of the passages in which Berlioz most nearly surrendered his individuality and conformed with the conventions of his day—the brigands in *Harold in Italy*,

Friar Lawrence's oration in *Roméo et Juliette*, and the entrance
of the Cardinal in *Benvenuto Cellini* are obvious examples. To
recognize the immeasurable superiority of Berlioz as a composer,
it is not necessary to ignore, or skate over, his petty indebtedness
to his contemporaries; and Meyerbeer, great eclectic though he
was himself, exercised a remarkably strong and wide influence
on the dramatic music written between 1835 and 1880. We
have already caught Wagner cribbing from *Robert le Diable* in
The Flying Dutchman and Verdi probably unconsciously
repeating an effect from *Les Huguenots* in *Rigoletto*. Before
coming to the general nature of that influence I should like to
trace in rather more detail the scenes in which that influence is
most often found, and give some familiar examples.

The supernatural treated as a theatrical effect, the 'demonic'
or the 'angelic', the cathedral or the witches cavern, prompted a
Meyerbeerian stock response. The evil chatter of the woodwind,
the tremolando strings or timpani roll, sudden fortissimo brass
chords of the dominant seventh on the one hand. And on the
other the harp arpeggios, the organ in a blaze of what the
Germans call *Scheinkontrapunkt* (the perfect examples of which
are to be found in the prelude to *Les Huguenots* and the
coronation scene in *Le Prophète*) and 'celestial' hymn-like four-
part writing in the woodwind as in *Lohengrin*—it is all very
familiar. Do not let us bother with operas such as Halévy's *La
Juive* and Wagner's *Rienzi*, frank imitations of Meyerbeer. But
think of Elsa's wedding procession in *Lohengrin*, with Ortrud
and Telramund the melodramatic embodiment of evil; of the
celestial voice at the end of Act III of *Don Carlos*, where the
whole scene is strongly marked by Meyerbeer's influence; of the
apotheoses in *Les Pêcheurs de Perles*, *Faust* and *Mireille*, all
modelled on the hymn from *Le Prophète*; of Gounod's
Mephistopheles and Ourrias (in *Mireille*); the temple scenes
evoked in *Les Pêcheurs de Perles* and actually presented in *Aïda*;
Berlioz's demons in *La Damnation de Faust*, Mussorgsky's
satanic Jesuit Rangoni in *Boris Godunov*, and the whole
confrontation of sacred and secular in Saint-Saëns's *Samson et
Dalila*—right down to Massenet's *Thaïs* of 1894, where the
climax of the ballet is a drawing-room waltz entitled 'La
Perdition', directly descended from the 'Valse infernale' of
Robert le Diable.

Another preoccupation of Meyerbeer's for which he found singularly apt expression was a certain slightly fatuous kind of courtly elegance—the charming frivolity of the Catholic noblemen and Marguerite de Navarre's ladies in the first act of *Les Huguenots*. The dotted rhythms, ben marcato, the cadences that suggest a bow or a hand-gesture, the elegant triplets and slightly affected staccato phrases, all combine to evoke the courtier. There is the entry of Raoul, the hero of *Les Huguenots*—a Protestant, it is true, but sadly tainted in his manners by the deplorable example of the Catholic court. Verdi turned to this kind of music for Gustav and Oscar in *Un Ballo in Maschera* and the Duke in *Rigoletto*, just as Bizet did for the Duke in *La Jolie Fille de Perth*; and minor French and Italian opera is full of examples.

Meyerbeer's influence is nowhere stronger on Verdi than in *Aïda*, not even, I think, in the operas specially written for Paris. Verdi knew that *Aïda* was to be a kind of occasional 'court' opera in the first instance; and the strong element of display suitable for the occasion prompted his return to the Meyerbeerian idiom. In one case he even borrowed an idea of Meyerbeer's. Ines's 'Adieu mon doux rivage', in which she says goodbye to Lisbon, was to recur to Verdi's mind when he made Aida think with nostalgia of her Ethiopian home.

Perhaps Amonasro is the most Meyerbeerian single character in *Aïda* and his 'Su dunque! sorgete egizie coorti!' could almost have been written by Meyerbeer at his best. How strongly this affinity was felt by contemporaries as shown by Vincent d'Indy's contemptuous dismissal of the whole of *Aïda* as 'that Meyerbeero-Wagneroid bore'.

When we come to consider the wider aspects of Meyerbeer's influence on his contemporaries and the next generation, we are confronted with the difficulty of distinguishing between the composer and the spirit of the age, the *Zeitgeist*, which he so perfectly embodied. The whole romantic aesthetic of shocking, dazzling, astonishing, is summed up in his work. Berlioz, who was by no means untouched by it himself, summed it up in a great tirade:

. . . high C's from every type of chest, bass drums, snare drums, organs, military bands, antique trumpets, tubas as big as locomotive smokestacks, bells, cannon, horses, cardinals under a canopy, emperors, queens in tiaras,

funerals, fêtes, weddings . . . jugglers, skaters, choirboys, censers, mon-
strances, crosses, banners, processions, orgies of priests and naked women,
the bull Apis and masses of oxen, screech-owls, bats, the five hundred
fiends of hell and what have you—the rocking of the heavens and the end
of the world, interspersed with a few dull cavatinas here and there and a
large claque thrown in.

What, then, are we to say of certain aspects of Berlioz's own
works—where we have a high C sharp, military bands, antique
cymbals, a cardinal under a canopy, a queen on a funeral pyre,
and the five hundred fiends of hell? Are these the 'influence' of
Meyerbeer or simply more moderate examples of the same
general aesthetic as Meyerbeer's? And what are we to say of
Liszt's virtuoso paraphrases, transcriptions, and studies in
'transcendental' virtuosity? They are exact instrumental parallels
to what Meyerbeer achieved in the theatre and were perhaps
inspired by this, as well as by the example of Paganini. We
happen to know what Chopin thought of *Le Prophète*, to which
he dragged himself a few weeks before he died, from a note in
the diary of Delacroix, who visited him the next day and spoke
of Chopin's 'horror of this rhapsody'. Yet the E major middle
section of the big A flat major Polonaise, with its ostinato octave
figure is a comparatively quiet and tasteful example of what
Liszt displayed with fewer inhibitions in *Funerailles*, and both
are quite in Meyerbeer's spirit.

Have Meyerbeer's latter-day champions—Joseph Conrad,
who preferred him to Wagner, Constant Lambert, and Bernard
van Dieren among musicians—really any case to make against
the oblivion which has in fact befallen his music? Is he in fact an
unjustly neglected composer? I should answer emphatically 'no'.
Despite some fine individual scenes and a handful of effective
rather than really powerful arias, his importance is entirely
historical. I have enjoyed a grand production of *Les Huguenots*,
but the production was as important as the music and that for us
today is a serious condemnation of any opera. I should not go so
far as Vaughan Williams and maintain that any music worth
performing is worth performing badly. But it is a fact of
everyday experience that the greatest music survives the most
extraordinary maltreatment; and as the essential quality of the
music itself deteriorates, so the quality of performance becomes
more and more important. We have only to remember what a

persuasive advocate Beecham was of second- and third-rate music. In the case of Meyerbeer, performance was everything. That is why he was willing to spend limitless sums, to consult the cleaners and the claque-members before deciding the final form of a scene; to hire the greatest singers and to insist on the most sumptuous productions—because without these things his works were no more than pasteboard façades, his characters are voices rather than three-dimensional human beings. Great men borrowed ideas from him and turned them to greater things; that was his true destiny and nothing more.

4

CARL MARIA VON WEBER[*]

No universally acknowledged 'great' composer of the last
century is represented in the modern repertory by so few works
as Weber; and few composers so narrowly escaped failure to
find themselves eventually counted among the immortals. All
composers, all artists indeed, pay a certain tribute to the
fashions of their day; and many cast their finest inspirations in
forms which never achieve permanent or universal validity. Bach,
Handel, Mozart, and Beethoven could write music as ephemeral
as the film music of today, and the music of some of the greatest
operatic composers—Monteverdi, Handel, Lully, or Gluck—
has frequently aged with the form in which they cast it. But
Weber is unique in having achieved a position among the great
although all his work is open to one or other of these main
objections. He wrote a mass of songs and chamber music,
almost all 'occasional' or designed for brilliant display; and the
best of himself he poured into his stage music and into two
operas, one of which is—in the words of Alfred Einstein—
'subject to ridicule on every stage except the German', while the
other is burdened with a libretto which would sink the music of
even the greatest genius. And yet Weber is not a figure of merely
historical or local importance. His best music has unique
qualities which are still recognized today, and universally
recognized, so that we ask ourselves what he might not have
achieved if his life had been different while it lasted and if it had
not ended at thirty-nine.

Carl Maria von Weber—the 'von' a pathetic fabrication of his
ludicrous father—was born sixteen years after Beethoven, in
1786, and he died a year earlier, in 1826. This father (uncle of
Mozart's wife) was a crazy, stage-struck beau with fantastic

---

[*] Martin Cooper, *Ideas and Music* (London, 1965), pp. 57–65.

ideas of his social status, who insisted on dragging his family from one petty German court to another and his youngest son, Carl, from teacher to teacher, in the hope of turning him into an infant prodigy. As if this unsettled life were not enough to ruin the young Weber, his father's plans went near to succeeding and Carl Maria found himself a theatrical conductor at eighteen, and something very near a hack composer. His musical career was variegated with attempts to make a fortune by lithography (which nearly cost him his life when he drank nitric acid by mistake) and by a spell as comptroller of the household to a dissolute duke at the court of his even more dissolute brother, the King of Württemberg. This ended in imprisonment and then banishment and a return to music as a profession, to restless journeyings up and down Germany and eventually to conductor-ships at Prague and, for the last eight years of his life, at Dresden.

This episodic, flashy existence was that of a hundred minor composers of the late eighteenth and early nineteenth century. But in Weber's case a kind of glamour was cast over it by his personality. It was not only his romantic excesses in wine and women, his poisoning, his imprisonment, his consumption, and early death that made him an early type of the new 'romantic' artist of the nineteenth century. Unlike the little court-composers and theatrical conductors of the eighteenth century, whom he resembled in the exterior course of his life, Weber was never content to be simply a musical tradesman, a craftsman. Indeed an absence of solid craftsmanship from his work earned him the contempt or disapproval of Beethoven and Schubert; and it accounts for Grillparzer's blank misunderstanding of him and Goethe's snobbish disregard. But, in exchange, Weber's deter-mination to be something more than a mere musician, to be a poet in sounds as well as a literary artist and a German patriot too, gives the best of his music a penetrating intensity and a freshness whose bloom never faded even when its novelty passed. It was this quality which caused Berlioz to idolize Weber and made first his old friend Meyerbeer and then, supremely, Wagner flatter his memory in the sincerest of all ways, by imitation.

The Abbé Vogler, Weber's chief teacher, has been described as 'one of the most devastating of musical humbugs'. He had been a great traveller, reaching not only Spain, Portugal, Greece, and Africa, but even, it is said, Armenia and Greenland. He made experiments in organ-building, developed a system of harmony based on acoustics, composed a Swedish opera and a Paradigm of the Ecclesiastical Modes (whatever that may mean), and wrote a small library of theoretical books. One attributed to him (and certainly its title suggests the author) is an *Ästhetisch-kritische Zergliederung des wesentlich vierstimmigen Singesatzes des vom Knecht in Musik gesetzten ersten Psalms*— or 'An aesthetic-critical dissection of the real four-part writing in Knecht's setting of Psalm I'. His appearance was described as that of 'a large, fat ape'. Not, in fact, a teacher to inspire confidence, possibly in great part a charlatan; and yet two of the pupils of his old age at Darmstadt, Weber and Meyerbeer, made operatic history, and cherished the kindliest and most reverential memories of their old master.

It was as a German patriot outraged by Napoleon's defeat and humiliation of his country that Weber first became known to the general public. Beside being something of a virtuoso as a pianist, Weber played the guitar and accompanied himself in his own compositions. Many of these were Italian canzonettas, and Weber's singing was one of his social assets rather than a serious musical activity. But when, after the dark days of Napoleon's victories, he set to music young Theodor Körner's patriotic songs, Weber's fame spread beyond musical circles. Körner was killed in the battle of Gadebusch at the age of twenty-two and his name and verses were thus lent something of the heroic glamour which attached to those of Julian Grenfell or Patrick Shaw-Stewart after the 1914–18 War. Weber's music in these *Lyre and Sword* songs looks back to the songs of Gluck rather than forward to those of Schubert; but his *Prayer during Battle* and *Lützow's Wild Hunt* achieved a popularity with successive generations of young Germans and it was to copy out the *Hunt* that Wagner bought his first music paper. The folk-song element—the other vein of originality in Weber's songs—found its fullest expression, of course, in *Der Freischütz* and in the songs it is still a quaint flavour rather than a deliberately pursued aesthetic policy.

Again and again we find this primacy of the theatrical in everything that Weber wrote. It is more than the expression of a dramatic temperament. The theatre and theatrical life were in Weber's blood, and it is difficult to say whether his unfaltering choice of the 'brilliant' and the 'effective' was conscious or, as I think more probable, instinctive and natural, at least by what we call 'second' nature. Meyerbeer we know to have been quite deliberate in his theatrical effects, but it is possible that both he and Weber were not simply fulfilling a demand of the fashion of the age but were also carrying on something which they had learned, perhaps by example rather than precept, from old Vogler, whose whole existence smacked of the theatre. The very forms of Weber's instrumental music often reflect the flashy, episodic nature of the talent which he expended upon them. There are too many 'grand' variations or pot-pourris; rondos and romances are too often Hungarian or Sicilian; moments are 'capricious' and polonaises unfailingly 'brilliant'; while empty-headed variations (exclusively melodic) and concertos for individual players are in disproportionate abundance.

Among all this occasional or 'social' music a few pieces stand out and have survived, but each of them bears serious blemishes. Take the four piano sonatas. In an autobiographical sketch Weber speaks of 'these accursed keyboard fingers which, by dint of everlasting practice, at last acquire a kind of independence, a wilful reason of their own and are unconscious tyrants and despots of our creative forces'. And so we find them. Again and again the thread of genuine musical thought and feeling is interrupted by pure finger-juggling, rocketing scale or arpeggio figures, lavishly ornamented tonic-dominant see-saw or 'brilliant' dotted rhythms, supporting a poor, conventionally operatic melody. Sometimes extraneous interest is provided, rather as in Meyerbeer's operas where each act has a revue-like sequence of 'novelties'. Thus the last movement of Weber's First Piano Sonata is a Moto perpetuo, the Second has a Menuetto capriccioso and the Fourth a Tarantella finale. Compare even the best—No. 2 in A flat—not with a Beethoven piano sonata, but with Schubert, and though Weber has poetic ideas, it is impossible not to feel the inferiority of his whole approach—glittering, superficial, and 'social', the *style galant* of the restored monarchies of Europe, a Congress-of-Vienna style. Even in the *Concertstück* the deeply expressive opening gives place first to the rather too easily raised storm of the Allegro passionato, with its interminable dominant seventh arpeggios; and then in the interests of the 'programme' first to the *naïvetés* of the C major March and finally to the empty finger-music of the Presto assai.

The *Concertstück* was one of the youthful Liszt's earliest battlehorses and Mendelssohn included it (with the *Invitation to the Dance* and the C major Polonaise) when he played to Goethe on 24 May 1830. How vivid an impression it made on Mendelssohn's own piano style can be seen from two works, both dating from 1832—the G minor Piano Concerto and the *Capriccio brillante* for piano and orchestra. Schumann showed in *Carnaval* that he knew the Menuetto capriccioso of the First Piano Sonata and, in general, it may be said that Weber's piano style formed the foundation of much of the 'brilliant' piano writing of the next generation. But, whereas Weber too often exploited brilliance for its own sake, Schumann and Liszt made it the vehicle of a new poetry and only Mendelssohn's lesser piano works continue the Weber tradition unaltered.

The *Invitation to the Dance*, far less essentially pianistic in character, is really a miniature symphonic poem (Weber himself provided a very nearly phrase-to-phrase programme) and it has lived in Berlioz's orchestration rather than in Weber's piano original. Its origin lay in one of those polite musical parlour games with which Weber amused his dilettante patrons—sitting at the piano and improvising musical illustrations to scenes or tales suggested to him by members of his drawing-room audience. In fact the *Invitation to the Dance* is the first of a series of 'apotheoses of the valse' which ended (we may now be permitted to hope) with Ravel's *La Valse* almost exactly a hundred years later.

To look for Weber's true physiognomy in his instrumental music is tantamount to looking for a poet to reveal himself in the small talk, however brilliant and occasionally profound, with which he may grace a cocktail party. If it is in his poems that we look for the poet, it is in his operas that we must look for Weber. It was there, in the opera, that every circumstance of his life and every trait of his character, inherited or acquired, would lead us to expect him to reveal himself. And yet, even here, what a disappointment at first sight! Of ten operatic embryos the first two have disappeared (Weber once said to

Schubert that a composer's earliest operas were like puppies: they should be drowned) while the third and fourth exist only in isolated numbers. In *Silvana* there are only occasional hints of the mature Weber's style and *Abu Hassan*, light-handed and witty as it is, still belongs essentially to the eighteenth century. *Die drei Pintos* exists only in fragments . . . All that remains is *Der Freischütz*, a single masterpiece; *Euryanthe*—potentially Weber's greatest music but indissolubly wedded to a nonsensical libretto; and *Oberon*, a hastily written work, a cross between romantic rescue-opera and fairy play, only remembered for its magnificent overture which is perhaps Weber's greatest single achievement, some fairy music and one big dramatic scene for soprano. It is really with this ill-assorted and patchy baggage that Weber has won immortality.

The historical importance of *Der Freischütz* in the development of German opera can hardly be exaggerated. Here, out of French melodrama and romance, Italian bel canto and German folk-song Weber succeeded in creating what Mozart had adumbrated in *The Magic Flute* and Beethoven (using the same ingredients) had hardly bothered to achieve in *Fidelio*—a full-scale German opera, with the same direct appeal to national sentiment as Verdi, Glinka, and Smetana were to sound in their different countries. The spoken dialogue and the plain humour of the old *Singspiel* are still there; and so are the stock characters—a timid, modest Virtue wooed by Youth, handsome but weak and led to the verge of ruin by fair-spoken Villainy; the Flighty Soubrette, the Pious Hermit and the Just Prince. But surrounding them, as never before, and casting a warm, new light over their familiar features is the atmosphere of the unmistakably German countryside, suggested by Weber not only in dances and choruses originating in German folk music but in the romantic forthrightness and simplicity of the characters. Each of these reflects in a different way that idealizing of the national character which every nation loves to find in its popular art. For *Der Freischütz* was immediately, as it has remained, a great popular work. It was thinking of *Der Freischütz* that Wagner, addressing Weber's spirit, cried: 'Today the Briton does you justice, the Frenchman admires you, but only the German can love you; you are his, a beautiful day in his life, a fragment of his heart.' Or a latter-day nationalist like

Hans Pfitzner could compare Weber's 'national mission' with that of Luther or Bismarck. We may leave such extravagant outbursts to those whom they may concern, among whom Weber is not to be found. Non-Germans may miss some of the overtones of Weber's music, but they will find the same melodic power, the same freshness of orchestral colour, the same magic rehabilitation of familiar forms as delighted Berlioz. With Berlioz we may agree that Weber's supreme gift is instrumental rather than vocal and that where, for example, he entrusts a melody on one occasion to a member of the orchestra (most probably his favourite clarinet) and on another to the voice, it is the instrumental singer whom Weber favours. To take a single example, it is the clarinet in the overture that we remember, not Max in the Wolf's glen, when we hear that most magical of melodies rising over the boiling tremolando of the strings.

In *Der Freischütz* Weber found the only coherent and truly musical libretto of his life. Its naïve horrors and naïver pleasures are as real and as valid within their own universe as the *naïvetés* of Haydn's *Creation*. But in *Euryanthe* Weber was confronted with the wholly unreal cardboard figures of a fourth-rate lady novelist's imagination. Here are the absurdities of *Robert le Diable* and *Lohengrin* in all their unredeemed crudity, and in fact both Meyerbeer and Wagner used the music of Weber's *Euryanthe* as a quarry. If there is truth in the old quip that *Rienzi* is Meyerbeer's best opera, it might with equal truth be said that *Lohengrin* is Weber's. How Weber, the first of the literary-minded musicians of the nineteenth century and himself a writer, ever came to accept Helmine von Chézy's libretto is a mystery; and the fact that, hotch-potch as it is, *Euryanthe* contains some of Weber's finest music is a proof of how easily ignited, how deeply saturated with musical ideas he was. Medieval chivalry attracted him, as the gipsy and oriental worlds had attracted him, by its remoteness and also by the dramatic possibilities in the extravagant contrast between radiant virtue and blackest vice. There is no hint of symphonic thinking in the concerted numbers of *Euryanthe*, any more than in any other work of Weber; and yet it was the suggestion of a new balance between music and drama which so excited Wagner in this score. Take only the purely instrumental poetry of the mysterious muted passage in the overture, or the introduction to Act II, where the orchestra speaks—as later in Wagner—with greater eloquence than any singer.

Euryanthe served as a kind of trailer or prospectus for some of the most effective scenes in nineteenth-century opera—the finale of Act II, with the heroine unjustly accused and humiliated in a huge public scene, looks forward to *La Traviata* and *Otello*, through Meyerbeer and Halévy; Lysiart and Eglantine closely foreshadow Friedrich von Telramund and Ortrud; and the hunting horns in Act III contain a hint of *Tristan*. In *Oberon*, written when Weber was dying, the poetry of orchestral colour far outweighs any other element in the music. The fairy horn-call, the gossamer patterns of the woodwind, the distant trumpet fanfare, the warmth of the divided cellos, and the velvet softness of the clarinet melody alternating with the idealized theatrical bustle of the strings and their own passionate melody—all these in the overture steal the thunder of the rest of the piece. Even Rezia's 'Ocean, thou mighty monster' culminates in a melody which the strings have already sung far better than any soprano can sing it, for the simple reason that it is an instrumental, not a vocal melody. Rezia's greeting of the sunrise contains the germ of Brünnhilde's awakening at the end of *Siegfried*. We can even find the Sword motive and Isolde waving her scarf to Tristan, in this scene. In fact Weber stands behind the young Wagner, not only as a musical model but as an inspiration, a living reminder of the potential greatness of German opera.

Anyone insensible to the particular quality of Weber's early romantic charm might claim, with some show of justice, that he did nothing but what was better done by his successors—might reasonably counter *Der Freischütz* with *The Bartered Bride*, *Euryanthe* with *Lohengrin*, and the fairy music of *Oberon* with Mendelssohn's *A Midsummer Night's Dream*; the piano music with that of Schumann, Liszt, and Mendelssohn; and the songs with those of any of the great *Lieder* writers. But such a critic would show himself a poor judge of musical character, for Weber's melody and the colour and texture of his orchestral writing have unmistakable personality. It is this that ensures him his place among the immortals even if, in Alfred Einstein's words 'in the semicircle that holds the niches of Bach and

Handel, Haydn and Mozart, Beethoven and Schubert, there
should at least be distinctions in the size of the heads' and
Weber's be smaller than the rest. Yet, we do not love people for
the size of their heads and Weber's ill, eager face with its hectic,
melancholy smile will never lack friends and admirers.

5

SCRIABIN AND THE RUSSIAN RENAISSANCE[*]

The bubble reputations of an age often reveal its character more clearly than those which are formed more slowly, and prove in the event more durable. The mid-nineteenth century, for example, classed Meyerbeer with Beethoven and Michelangelo, and this, though it tells us little enough about Meyerbeer, tells us much about the mid-nineteenth century. Even great artists are often admired in their lifetimes for qualities which seem to later generations either non-existent or secondary. Their greatness, though felt instinctively by their contemporaries, is accounted for in terms of the fashionable philosophy of the day, which often proves inadequate.

Bubble reputations, on the other hand, attach to one of two classes of artist—those whose art merely satisfies the superficial fashionable taste of their age (Meyerbeer is the classical example) or those who appeal to a deeper, more enduring need but at a superficial level; and of those, Scriabin is almost a unique representative in musical history. Fifty years ago Scriabin's music was admired as that of a harmonic innovator and a new spiritual force. Today his harmonic 'innovations' appear as no more than the exploiting *ad nauseam* of a single chord, while his mystical beliefs are regarded as psychological fantasies of purely clinical interest. How are we to account for this complete reversal of opinion?

The generation in which Scriabin grew up was 'looking for a sign'—for some new extension of the language of music and some new revelation to take the place of Wagner's now familiar theodicy. Some found the new Messiah in Debussy, some in Richard Strauss, a few in Mahler; but in Russia intellectual and artistic circumstances were so different from those in western

[*] BBC 6 October 1957.

Europe than none of these qualified for the Messianic role. Pressure had been steadily accumulating throughout the second half of the nineteenth century behind the movement of political and social reform and intellectual emancipation, and by the 1890s it had reached a strength and a density at which the smallest spark would, it seemed, cause an explosion. With all social and political activity denied them, thinkers and artists turned with hysterical intensity to irrational, mystical, and unnaturally 'other-worldly' fields of interest.

Here the genuine was inextricably confused with a charlatanism which was often unconscious. Even as early as the late 1870s, Dostoyevsky, working on *The Brothers Karamazov*, had been deeply impressed by the writings of a certain Fyodorov who believed it to be the task of philosophy 'to raise the dead collectively, end childbearing and usher in the transfiguration of life here on earth'. Solovyov, a really distinguished religious thinker and poet on whom Dostoyevsky modelled the character of Alyosha Karamazov, had apocalyptic dreams of the end of the world and was vouchsafed a vision of the Divine Wisdom in the reading-room of the British Museum. In fact this apocalyptic quality—this feeling that the end was at hand—was the underlying characteristic of all Russian thought from the late 1890s onward. Solovyov, who died in 1900, gave expression to this feeling in a much quoted line—'the end is already near; the unexpected will soon be accomplished'. And other writers spoke of 'the feeling of sickness, alarm, catastrophe, and disruption which lay heavy on the last generations of Russian thinkers before the Revolution—the constant and wanton feeling of catastrophe evoked by an accumulation of indisputable facts'— 'the search for an integral way of life, for a single answer to all life's problems, some form of collective which would weld together the sundered fragments of Russian life.'

In this atmosphere the Messina earthquake of 1908 was seriously regarded by many as an apocalyptic portent. 'We know what the fragrant names of Calabria and Sicily mean,' wrote the poet Alexander Blok, 'but let us be silent and grow pale, knowing that if the ancient Scylla and Charybdis vanished from the earth, yet ahead of us and in the heart of the earth a more terrible Scylla and Charybdis await us. What can we do . . .? We can only put on mourning, celebrate our sorrow in the face of the catastrophe. The battleship lowers its flag to half-mast—as though it were a sign that the flag had been lowered in our own hearts. In the face of the raging elements the haughty flag of culture is lowered.'

In this tense and nervous atmosphere, every kind of new religion and superstition proliferated: pantheism, theosophy, anthroposophy, and the new semi-oriental gnostic philosophies of Gurdjiev and Uspensky, the erotic mysticism of Rozanov, Gorky's 'demotheism', and Merezhkovsky's belief in 'the secret three designed to be the nucleus of the new church of the Holy Spirit in which the mystery of the flesh awaited final manifestation'—a curious prototype of D. H. Lawrence's mystique of sensuality. There was much talk of 'the new religious consciousness' and the journal issued under Merezhkovsky's guidance was characteristically called first *The New Way* and later *Questions of Life*.

Perhaps the most characteristic of all was the equation of Christ and Dionysos, actually proclaimed in a book published in 1903 by the poet Vyacheslav Ivanov. The shadow of Nietzsche lay heavily over all these thinkers and poets who were united in their search of 'ecstasy for ecstasy's sake'. This was to be sought, according to Vyacheslav Ivanov, in 'symphonic culture' and non-acceptance of the world. In fact, the universal divorce between speculation and the reality implied in this 'non-acceptance of the world' reflected the tragic separation of culture from all practical, social, or political activity, and was eventually to infect the political world; so that even the political history of the years between the Russo-Japanese War and the 1917 Revolution often shows the same unreality and dream-like hysterical quality as the symbolist plays and gnostic manifestos of the previous decade.

It was in these years and against this background that Scriabin grew up and made his reputation, for it was a Russian reputation in the first place. Music, as often before, lagged behind the other arts and was still suffering from Wagnerian fever when the Symbolists and the religious-literary sects had moved to a correspondingly later phase. At the end of a long life Rimsky-Korsakov, it is true, paid tribute to the prevailing 'spiritualism' of the day in his opera *The Invisible City of Kitezh and the maiden Fevronia*, a blend of naturistic pantheism with

Orthodox symbolism. But Rachmaninov continued the Tchai-
kovskian tradition of romantic introspection, penetrated with
the sense of impending doom and the nostalgia for a non-
existent past, which was part of the very atmosphere that he
breathed in Russia.

It was in the late 1890s that Rimsky-Korsakov wrote of 'that
star of the first magnitude newly arisen in Moscow—the
somewhat warped, posing, and self-opinionated Alexander
Nikolaevich Scriabin'. Born in 1872, Scriabin was indeed
exceptionally and precociously gifted but in many ways the
characteristic product of a too exclusively feminine upbringing
by the aunt and grandmother who took the place of the mother
who died in his early childhood. He was a dandy in the
aristocratic Cadet Corps before he entered the Moscow Conservatory to study composition with Taneiev and the piano with
Safonov, winning a gold medal as a pianist in 1892. His earliest
published compositions, dating from the late 1880s, show a
kinship with Chopin so marked and so all-pervading as to
present a case of something like spiritual identification. They are
miniatures of a 'salon' type, extremely elegant in form and
facture but already distinguished by a heavy, disturbing, and
individual scent very different from the characteristically
innocent and flowery scent of Chopin's smaller pieces.

The markedly 'indoor' character of the early music of
Scriabin, with its literary and urban atmosphere, is never
stronger than in the pieces of strongly erotic character, whose
inspiration is plainly the boudoir rather than the salon. The
erotic element, which remained one of the poles of Scriabin's
art, already foreshadowed a kinship with Liszt that was to grow
stronger as the composer grew increasingly discontented with
the concept of music as a self-sufficient art and reached out
vaguely towards a philosophical or religious meaning—or, in his
own words, to 'the *being* of which every work of art is only a
becoming'. The *Poème Satanique* of 1903 not only echoes
Liszt's *Malédiction* and *Mephisto* pieces but shows Scriabin
committing himself to the magical view of art which, as we have
seen, prevailed in Russian intellectual circles of the day. He
numbered among his friends Prince Sergei Trubetskoi, one of
two philosopher brothers and the friend and protector of
Solovyov; Merezhkovsky and his wife Zinaïda Gippius; and

Vyacheslav Ivanov at whose flat in St Petersburg—significantly called 'The Tower'—the young Berdyaev attended an apparently very mild attempt to revive the 'Dionysian mysteries'.

Already, in 1900, Scriabin's First Symphony had shown a new taste for the grandiose and the religious. It is in six movements, the last of which is choral and consists of a Hymn to Art as a 'wonderful image of the Godhead'. With the Fourth of the Piano Sonatas, written in 1903, he has begun to throw off the drawing-room elegance. Here we have already the fragmentation of melody, the caresses, sudden winged phrases, and ecstatic trills that herald his mature manner.

Despite the manifest echoes of Wagner in that music, Scriabin had already rejected the idea of an opera, on the grounds that it could only be the 'representation of a dramatic action, not the act itself'; and this idea of a musical performance as a magical rite, a liturgical incantation, the calling to life of hidden cosmic forces, completely dominated him from now onwards. It is explicit in the inscription over the visionary Fifth Piano Sonata: 'I call you forth to life, hidden influences, sunk in the obscure depths of the Creative Spirit, timid germs of life, I bring you boldness!'

Hitherto Scriabin, for all his desultory reading in philosophy, had been a mystic without a theology. The movements of the Third Symphony, written in 1903–4, bore significant but still vague titles—'Struggles', 'Delights' and 'The Divine Game'—this last an echo perhaps of Hegel's 'endless play of Love with itself'. By this time he had taken a positive and final dislike to the music of all other composers and was thus, both as artist and thinker, enclosed in a completely solipsistic world. The exact date of his first interest in theosophy seems to be uncertain, but he first became acquainted with the writings of Blavatsky and Annie Besant during a visit to Paris in 1907, and it was then that he developed something approximating to a philosophy of art. From now onwards he saw himself as the Messiah, destined, as he believed, to bring about the Final Act by which Spirit was to redeem Matter; and a great liturgical rite, in which all the arts were to play a part, was to usher in a new era. We can find in his own writings as clear a statement as such beliefs permit—'an ocean of cosmic love encloses the world', he writes, 'and in the intoxicated waves of this ocean of bliss is felt the approach of the Final Act—the act of union between the Male-Creator and the Woman-World'.

This sexual imagery always persisted and even so late a work as the Seventh Piano Sonata, written in 1911–12 and particularly valued by the composer, contains—according to Leonid Sabaneiev, the composer's close friend and disciple—a naïve and rather crude erotic 'programme'. Unlike the vast majority of mystical thinkers, Scriabin always allots himself the male role in the mystic marriages and acts of union, of which he writes. 'Oh! I would I could possess the world as I possess a woman', he exclaimed, and his mystical cosmology is constructed round his own creative personality. In the first process of creation, Spirit and Matter were one: they separate only in order to create the world and then unite once again.

Scriabin spoke of this first phase of creation as 'creative agony' or 'lust for life', and it is followed by a second phase, the process of dematerialization. 'The world glitters with the imprint of the Creator Spirit's beauty,' he wrote, 'but at the same time it moves further and further away from the Creator, diffusing itself in innumerable protean phenomena.' The desire of the world to be freed from the bonds of matter—and we are immediately reminded of St Paul's 'the whole creation groaneth and travaileth'—could only, he believed, be fulfilled by means of art, or a synthesis of the arts, in the hands of a Messiah. Like many of his contemporaries in Russia, he attached particular importance to India, the home of 'ancient esoteric knowledge', and even planned to buy a site there for the temple in which the Final Mystery was to be accomplished. All his later works he regarded as sketches for this 'Preliminary Act', fragments of whose literary text were among the papers found at his death in 1915.

The nearest he approached to his ideal was probably the *Poem of Fire* or *Prometheus* for orchestra, piano, and *clavier à lumières*. This 'light-machine' was a first attempt to achieve a synthesis of the arts and a 'counterpoint of the senses' which Scriabin believed essential to his composite form. He dreamed of 'a musical phrase ending in a scent, a chord that resolves into a colour, a melodic line whose climax becomes a caress'. And he undoubtedly saw himself as Prometheus (an extension of his Satanic fancies), the Free Redeemer rescuing the world of matter by the power of spirituality and 'ecstasy'. The music itself makes a hypnotic effect by its repeated insistence on small cell-like phrases, harmonies built on superimposed fourths, and the profusion of trills. The French marks of expression are a further guide—dark, threatening, strange, charmed, limpid, defiant, stormy, bellicose, like a winged caress, with an intense desire, like a shout, glittering, ecstatic, are a few of them.

Listening to this unfolding of an *idée fixe*, it is easy to understand that its appeal was largely to those who already felt the attraction of Scriabin's personality and were moved by the strength of his personal conviction; and most of all to those Russian audiences to whom the spiritual background of the music, so completely alien today, was already familiar. The nearest parallel at all familiar today is to be found in the gnostic writings of Gurdjiev. Like Scriabin planning his Final Act, Gurdjiev spent years on the plans for a vast mystical ballet, believed in the possibility of a world-language, and in the magical, theurgic power of art. Scriabin compared his last two piano sonatas to magic rites, the Ninth a black Mass and the Tenth 'white' or beneficent magic; and in both, the fragmentation of musical material reaches nearly the same point as Schoenberg was reaching at the same time in his Opus 11 piano pieces. We can see the final point reached by Scriabin in the Ninth Sonata, written in 1912–13.

What are we to make of Scriabin? It is easy to dismiss him as a spoiled talent, the composer of the youthful piano pieces in which the spirit of Chopin is wonderfully revived and continued, but later spoiled by crazy mystical notions and an obsession which brought him to the verge of lunacy. In fact Scriabin's later piano sonatas, however esoteric their musical content, remain interesting from the purely pianistic point of view; and *Prometheus* and the *Poem of Ecstasy* will retain their place in the history of the orchestra—a place with, say, Schoenberg's *Pelleas und Melisande* and Strauss's *Domestic Symphony*. But today Scriabin's career is chiefly interesting as a warning, a kind of cautionary tale of the man who lost his ability to distinguish between dream and reality because he lost contact with humanity.

This, as we have seen, was the great weakness of the Russian artistic and intellectual 'renaissance' at the opening of the present century and its explanation lies in the political, social, and religious life of Russia. The Revolution shattered the dream and blew up the ivory towers; but it has hitherto imposed on its artists a view of 'reality' almost as one-sided as that of the aesthetes and mystics. For the artist cannot restrict his human interests and sympathies to any social class or imagined category. Even the exclusive concern of the French classical dramatists with royal or princely personages is less restricting in fact than the Russian artist's obligatory concern with proletarian characters. It is an unimportant convention that Racine's Phèdre is a queen, for she is first and foremost a human being, whereas the creations of Soviet artists too often conceal beneath their party or proletarian status not so much a human personality as an ideological blueprint.

6

SERGEI PROKOFIEV[*]

According to many western critics Prokofiev's career was a slow degeneration from the brilliant and ruthless young musical anarchist of pre-revolutionary days, through a period as 'playboy of the western world' during the 1920s, to the respectability of the Soviet laureateship and the unconscious philistinism of his latter years. The official Russian portrait, on the other hand, is of a musical 'prodigal' who left the fatherland to go whoring after the strange gods of the West and returned in middle life to his duty, to die in the odour of Soviet sanctity.

I believe that we may hope to understand Prokofiev if we compare him with his contemporary, compatriot, but absolute antithesis—Stravinsky. Stravinsky and Prokofiev have pursued diametrically opposite paths of development as artists, yet both have travelled the same psychological 'inner circle'—Stravinsky travelling, as it were, from east to west and Prokofiev from west to east. The differences that divided them were by no means merely temperamental, though those were certainly great. Entering the Russian musical world ten years later than Stravinsky, Prokofiev found that an impasse had been reached— a point at which established composers had exhausted the traditional language but still feared to embark on new experiments.

Prokofiev's hostility to the Russian musical establishment of his youth had the character of an instinctive recoil, immediate and prompted by no intellectual reasoning. In his recent *Conversations* Stravinsky describes Prokofiev as 'the contrary of a musical thinker', and goes on to make fun of his commonplace mind and poor general culture. This is of course the perennial jibe of the intellectual musician against the musical 'natural'. It was Prokofiev's musician's instinct, not any conscious intel-

[*] Martin Cooper. *Ideas and Music* (London, 1965), pp. 135–41.

lectual choice, that drew him from the very beginning towards Western music. Indeed the seeds of neo-classicism were perhaps first sown during one of those evenings in December 1906 when Max Reger visited St Petersburg and gave a concert of his own works. Prokofiev was then a boy of fifteen, and the impression that Reger's music made on him can be seen clearly in the *Ten Pieces* for piano that he wrote during the next eight years.

If we consider such music against the background of Rachmaninov or Scriabin, its full originality appears. Here are fresh air, a clean palette, and no more souls or symbols, but also no theories of regenerating music by a return to the past, no intellectual justifications or *raffinement*. Prokofiev was indeed a self-conscious 'barbarian' in those days, and almost at the same time as the Regerish neo-classical pieces he wrote the *Toccata* whose persistent drumming rhythm and thick, graceless harmonic chunks must have sounded crude indeed in 1913.

The two chief centres of musical modernism in Russia during the years immediately before the Revolution were the Evenings of Contemporary Music in St Petersburg and the Moscow magazine *Muzyka*. Both these institutions gave the young Prokofiev their fullest support and brought him into contact with the world of Diaghilev's magazine *Mir Iskusstva*, 'The World of Art'. This represented the reaction against what was then the Tolstoyan, and is more familiar to us as the Soviet, view of the artist as a moral agent with grave social responsibilities. When Diaghilev proclaimed that 'the only way to ensure progress and effectively to combat routine in the arts is to follow contemporary movements in Western Europe', he was declaring himself the last of the maximalist Westernisers. How seriously Prokofiev was affected by any theories is doubtful (he always hated theorizing), but there is no doubt that he felt the influence of Diaghilev's trump-composer, Stravinsky, and showed it clearly in his *Scythian Suite*, close to *Le Sacre du Printemps* in conception and often is sonority, and in the ballet *Chut*. Moreover, although he always remained on the outskirts of Diaghilev's personal circle, which regarded him as embarrassingly uncultured, what we may call a 'World of Art' strain persisted in his choice of subjects until the late 1920s. We can see it in his preference for the poet Balmont, who provided him with the texts of a number of songs and the cantata *Seven of Them*; for

Bryusov's *The Fiery Angel* with its magic and its religious mania; in the brittle, fantastic world of Gozzi's *The Love for Three Oranges*, and the sophisticated, abstract, and modernistic ballet *Le Pas d'Acier*.

Prokofiev's initial attitude to the revolution was seemingly one of incomprehensible indifference. Music was in fact his sole interest, and he shared the common belief that a revolution which aimed at clearing the ground for a new society must automatically favour artistic innovation. When in 1918 he was disappointed in this belief, he left Russia without hard feelings and without cutting the ties that bound him to his past. For the next fourteen years he led the life of a travelling virtuoso pianist, with intervals dedicated exclusively to composition.

A list of the works completed or composed during these years reflects Prokofiev's instinctive preferences while he was under no sense of obligation, concerned only with satisfying his own creative impulses and making a living. In the first place we have three operas: *The Gambler, The Love for Three Oranges*, and *The Fiery Angel*, and four ballets: *Chut, Le Pas d'Acier, The Prodigal Son*, and *Sur le Borysthène*. Closely related to these are the Third and Fourth Symphonies, based respectively on material of *The Fiery Angel* (never performed at the time) and *The Prodigal Son*. Since Prokofiev was making his livelihood as a pianist, it is strange that he wrote only one piano sonata and one concerto during these years. A handful of chamber works completes the list. None of this music brought Prokofiev any real and lasting success.

He was not, in fact, a lovable personality; and his music aroused in the Western European and North American public an unsatisfactory blend of admiration and even repulsion—'the football pianist', they called him. Prokofiev himself declared that his chief interest at this period was 'the search for novelty and the breaking of tradition', and it does not need a Soviet aesthetician to find this a negative and incomplete ambition for a mature artist. The first sign of an approaching crisis in Prokofiev's artistic development was a falling-off, not so much in the quality as in the quantity of his work. Like all musical 'naturals', he had always written with extreme facility and seemed never at a loss for a musical idea; but during the latter part of the 1920s we find him repeatedly rewriting or arranging unsuccessful works rather than composing new ones. So that by the time he decided to return to Russia his output was in fact reduced to a trickle. His music was not wanted in Western Europe, and he needed the stimulus of performance, needed to put his gifts at the service of some welcoming and approving body. His thoughts returned to what was now a new public, his own people. In 1930 he was approaching forty and found himself at an age and in a position which send many prodigal sons back to their fathers, or their fatherlands. In 1933 he tried to explain to Serge Moreux what was now a firm decision to return to live in Russia:

I must see a real winter and a real spring again, I must hear Russian spoken round me and talk to people who are close to me . . . whose songs are my songs. Here I am restless and I'm in danger of becoming an academic.

This was a strange word and one that he used perhaps already in the Soviet sense of 'rootless' and 'formalistic'.

Was his return what Stravinsky calls it, 'simply a sacrifice to the bitch goddess', a confession of failure and a desperate attempt to win success by a recantation? It was not as simple as that, though I believe that Stravinsky's religious metaphor is apt, and that Prokofiev's return to Russia was in a sense a religious act, an act of faith and an attempt to 'save his soul', which he felt parched and withered in the fashionable avant-garde musical circles of Western Europe. Certainly his fertility returned immediately and never left him again, though we can find evidence of his filling periods of diminished creative power by copious arrangements of his own music. In the twenty years of working life that were left to him—the whole of the second half of his life as a composer—Prokofiev wrote two-thirds of his whole output. Much of this music consists in frank trivialities, but there was also much serious and, in a new way, characteristic music.

'My lyrical gift,' he once explained, 'was slow to develop because it was so little appreciated', and it is true that he was accepted in Western Europe as the extrovert, sarcastic, irreverent playboy of the contemporary movement, the monkey whose tricks, as he got older, gradually ceased to amuse. Now, back in Russia, he was asked to develop his lyrical gift, and according to a formula that he had in the past repeatedly mocked—broad, immediately intelligible (that is to say traditional) melody of a 'life-enhancing' type modelled either on Russian folk-song or the Russian popular classics—which meant Glinka, Rimsky-Korsakov, Mussorgsky (though with reservations), and Tchaikovsky. The mature Prokofiev was incapable of imitation or pastiche of any kind. He was absolutely without Stravinsky's strong historical sense, the feeling for style and period. Although a self-borrower and arranger on a scale which outdid even Handel, he had none of the sophisticated musician's interest in recomposing other men's music or in recreating any past style.

The lyrical vein, which had lain for the most part dormant in him hitherto and was now consciously developed, was genuinely popular and unmistakably national, as we can see from two works composed before he was thirty. The first of these was a setting for voice and piano of Hans Andersen's *The Ugly Duckling*. This was written in 1913, at the height of his ruthless and brilliant youth when he was generally supposed to be devoid of all 'humanity'. It was a favourite with Gorky, who used to say that Prokofiev himself was the ugly duckling, pecked by critics and older musicians and spurned by the public but, in fact, a swan. Prokofiev's graphic music for the duckling's rearing, education, and first acquaintance with unhappiness vividly recalls Mussorgsky, with its brusque changes of rhythm and tempo and its folk-ish air, just as it clearly foreshadows the duck of *Peter and the Wolf*.

The other work in which we can find clear evidence of the simple, popular lyrical style that Prokofiev developed so assiduously during the last twenty years of his life is a collection of four piano pieces written in 1918 and bearing the significant title *Tales of an Old Grandmother*. This popular traditional lyrical style was developed by Prokofiev after his return to Russia in an enormous number of songs and small instrumental pieces. The success of his large works, too, with public and critics, seems to have been in exact proportion to their national and traditional lyrical content. In these works Prokofiev either silenced his sarcastic wit and his musician's instinct to explore new fields, or he skilfully discovered some external dramatic reason for indulging them—as in the *Alexander Nevsky* cantata, where the enemies of Russia are depicted in music that would not otherwise have won official approval.

On the other hand there were works in which the composer imposed less restraint on his imagination and showed that the old violent, bitter-tongued, mocking Prokofiev was by no means dead: these included the Sixth Symphony, the First Cello Concerto in its original form, and the Seventh Piano Sonata. These were apparently unanimously declared 'negative' and regarded as unfortunate lapses into the bad habits acquired during the years of expatriation and Western contamination.

It would be false, I believe, to suppose that Prokofiev took refuge in some elaborate system of 'double think'—a kind of self-induced musical schizophrenia. The whole interest of his life was composition and he was fundamentally indifferent to doctrinaires of all kinds, the musical doctrinaires of the West and the politicians in Russia. He discovered, though, that he could satisfy the politicians without offending any deep conviction of his own, and thereby win a position in which he could be sure of having all his works performed and at the same time raise the level of Soviet music.

I should not expect Prokofiev to be remembered by the works such as the *Scythian Suite* or *Chut*, in which he was competing with Stravinsky or, on the other hand, by *War and Peace* or *Semyon Kotko*, in which he was merely doing more successfully what other Soviet composers were also doing—composing to a formula accepted for other than musical reasons. *The Love for Three Oranges* and *The Fiery Angel*, and possibly *The Betrothal in a Monastery*, are, on the other hand, unique both in conception and execution. As a symphonist Prokofiev showed a characteristically Russian inclination to confect symphonic suites rather than symphonies originally conceived as such; and he is more likely to be remembered by his ballet music and his concertos, a form far better suited to his extrovert temperament, while the nine sonatas and more than one hundred and twenty pieces that he wrote for the piano constitute a body of keyboard music unique in the twentieth century.

There is a considerable amount of repetition in his music: rhythmic figures, harmonic progressions, and melodic shapes recur from work to work with only slight variation. But at his best his music has a character unmistakably its own. In judging the man and his work many people have been inclined to use criteria which do not apply to his psychological and aesthetic type, quarrelling (as it were) with a Siberian crab-apple for not producing peaches. Prokofiev remains one of the first half-dozen composers of the twentieth century, and until Benjamin Britten he was almost alone in demonstrating that the gap between the consciously 'contemporary' composer and the general public was not unbridgeable.

7

GOUNOD'S INFLUENCE
ON FRENCH MUSIC*

Charles Gounod has probably succeeded Mendelssohn in the mind of the semi-educated musical world as the representative of all that twentieth-century music does not stand for. The younger generation may or may not, in its heart of hearts, care for *Faust*; but *Faust* is the only work of Gounod's which it knows, and it is not difficult to find material for contemptuous jokes in the libretto and the music. (It never has been difficult. In the 1880s, when *Faust* was a little more than twenty years old, Saint-Saëns and Chabrier used to do a party performance of the church scene, Saint-Saëns singing the part of Marguerite.) An older generation in England remembers the days when *La Rédemption* stood almost on an equal footing with *Messiah* and *Elijah*; and it is impossible for anyone who has ever been a regular attendant at Anglican services not to have something of Gounod's religious idiom so deeply engrained in his musical consciousness as to be almost second nature. Gounod's evil reputation reposes, in fact, on a vast ignorance—on a shamefaced affection, often hotly denied, for *Faust*; on a healthy dislike of the mental and emotional furniture of his weaker disciples, Sir John Stainer and John Bacchus Dykes ('Our blest Redeemer, ere He breathed', 'Holy! holy! holy!') and on a rather exaggerated horror of the 'Ave Maria' arranged by Gounod as a descant to the first prelude of Bach's *Forty-Eight Preludes and Fugues*. Probably not one in a hundred English music-lovers realizes that Gounod was in his day a revolutionary, a stern protestant against the debased musical taste of his age, a fanatical admirer of Bach and Palestrina, and one of the heralds of the French musical Renaissance which flowered so quickly and so abundantly in the years after the Franco-Prussian war of 1870.

* Martin Cooper, *Ideas and Music* (London, 1965), pp. 142–52.

Born in 1818, Gounod grew up in the years of clerical and monarchist reaction which followed the defeat of Napoleon, the end of the First Empire, and the restoration of the Bourbons. The atmosphere of those early years combined with Gounod's naturally emotional and idealistic temperament to make religion one of the most persistent and one of the deepest sources of inspiration throughout his life. At the Conservatoire he was accepted as a pupil as young as seven years old, studying with Lesueur, Reicha, and Halévy, who were probably the finest masters any young musician of the day could have found in Europe. Lesueur was the solitary representative of the old Gluck tradition in France, which had borne a single fruit in Spontini and then almost disappeared except in the mind of this single old eccentric, with his passion for the vast, 'antique' musical canvases of the Revolution, his interest in Greek, Roman and Hebrew music, and his inflexibly high principles, which he succeeded in handing on to another favourite pupil, Hector Berlioz. Reicha was an immensely learned contrapuntist, who had moved in Viennese musical circles, had known Beethoven personally and cherished a vast admiration for his works, which were still virtually unknown in France. Halévy was a brilliant and gifted theatrical composer, who himself produced at least one fine opera (*La Juive* 1835) and trained a whole generation of French composers, including Georges Bizet.

The combined influence of these three men had a permanent effect on Gounod's musical ideals to the end of his life; and when, in 1839, he won the Prix de Rome, it was as a fervent idealist and a determined opponent of the fashionable worship of Rossini, Donizetti, and Meyerbeer that he set out for the Villa Medici. In Rome he came under other formative influences which further moulded his outlook. Ingres was then head of the Villa Medici and his passion for music naturally inclined him to take a special interest in the musicians in his charge: Gounod had charm and enthusiasm and it was not long before he had become a personal friend of the director. Musically the most important of his Roman experiences were his friendships with two women, Pauline Viardot and Fanny Mendelssohn, the one a great singer and the other a fine pianist. Pauline Viardot had some of both the charm and the artistic ability of her more famous sister, 'La Malibran', the romantic ideal of a tragedienne, who inspired the whole of the literary *Jeune France* by her voice, her personality, and her tragically early death. The influence of Fanny Mendelssohn was entirely different. She was a cultured German woman who introduced Gounod to the works of Bach, Beethoven, and her own brother, to the classics of German literature and especially to Goethe, laying the foundation for that fine workmanship, that thoroughness and competence which were to distinguish all Gounod's compositions, even the most trivial. Finally, in that other sphere which he always felt instinctively to be the basis and foundation of his whole life and work, Gounod fell completely under the spell of the great Dominican preacher, Père Lacordaire, whose sermons caused a sensation in Rome between 1838–41. These three personalities— the two women, so utterly unlike yet both devoted to music, and the Dominican preacher—represent the three fundamental elements in Gounod's character, the three passions of his life: music, love and, transcending both, religion.

In 1842 Gounod decided to move on from Rome to Vienna, where he was to begin the German portion of his Prix de Rome years. He stayed there a year, hearing a great deal of music (including the works of Beethoven) and having two masses of his own performed; and then moved on to Berlin, where he saw Fanny and her husband, and finally to Leipzig, where he spent happy days with Mendelssohn himself and was introduced to the magnificent Gewandhaus orchestra, which was probably the best in Europe during the 1840s.

It was 1845 before Gounod was back in Paris, and he had been six years away from France. His first action was to ensure a livelihood by accepting the post of organist to the fathers of the Missions Etrangères in the Rue du Bac, where he shocked many of the congregation by introducing the works of Bach and Palestrina in place of the accepted theatrical vulgarities which passed as church music in the 1840s. But a decision was gradually being thrust on him, a choice which could not be indefinitely postponed. Lacordaire's preaching had fallen on fruitful ground, and the daily familiarity with the heroism and piety of the fathers in the Rue du Bac had its inevitable result. In 1847 Gounod decided to start reading for the priesthood and entered the seminary of Saint-Sulpice. He was not there long. Either the call of music and the world was too strong or, as he is said to have admitted himself, he feared the emotional intimacy which he did not feel himself strong enough to prevent arising between himself and the women who might bring their secrets and their troubles to him in the confessional. Whatever the exact reason may have been, Gounod renounced the idea of the priesthood and left Saint-Sulpice.

In 1851 his renewed friendship with Madame Viardot led to the composition of *Sapho*, his first opera; and in the same year his *Messe solennelle* was performed in London. His career as a composer had started; but it was by no means all roses. *Sapho* was praised by Berlioz, who nevertheless found it too fierce—'il faut avant tout qu'un musicien fasse de la musique', he wrote, reversing (as modern eyes see it) the roles of lion and lamb. One penetrating critic discovered the real inspiration of the music, which was the operatic ideal of Gluck, as Gounod must have come to know it from Lesueur. *Sapho* is truly dramatic music, firmly but finely drawn, with a charm and discretion, an absence of over-emphasis, and a charity of style which are among the greatest qualities of the French artistic genius. The incidental music to Emile Augier's *Ulysse*, which appeared the following year (1852), was as apt, as charming and as little successful with the general public as *Sapho*; and in despair Gounod, who had married in the meanwhile a daughter of the musician Zimmermann, turned to a more popular genre. In 1854 his *La Nonne sanglante* was produced at the Opéra. If Gounod had hoped for a repetition of the success of Meyerbeer's *Robert le Diable*, he was disappointed; for although his score was praised by both Berlioz and Théophile Gautier, it is inferior to *Sapho* and *Ulysse*, and certainly to the works of his own which were to follow in the immediate future. But he was making his mark. In 1855 Berlioz wrote of the musical life of Paris: 'à part . . . C. Saint-Saëns, un autre grand musicien de dix-neuf ans, et Gounod, qui vient de produire une très belle messe, je ne vois s'agiter que des éphémères au dessus de ce puant marais qui s'appelle Paris.'

In 1857 Gounod succumbed to a nervous breakdown, in which it was feared that he would lose his reason; but he recovered quickly and so completely that during the next three years, 1858–60, he wrote three of his finest works: *Le Médecin malgré lui*, *Faust* and *Philémon et Baucis*. The Théâtre Lyrique was the least conservative of the three theatres in Paris where a young French composer could hope to have his works performed, and it was there that *Faust* and *Philémon et Baucis* were given. *Le Médecin malgré lui* was a brilliant restatement of Molière's comedy, witty, quick-moving and alive; but both this and *Philémon et Baucis* were inevitably swamped in the glory of *Faust* and its more showy successors, *La Reine de Saba* (1862) and *Roméo et Juliette* (1867); nor were they fully appreciated until they were rediscovered and performed after the First War. Both are what might be called 'chamber' operas which lose a great deal of their charm and point if they are performed on too large a scale; both are thoroughly individual and thoroughly national in a way that *Faust*, *La Reine de Saba* and *Roméo et Juliette*, with their frequent overlay of Meyerbeerian glitter, could never be.

It is impossible to discuss *Faust* in detail here, but it is essential to realize, especially in this context, that the charm and the greatness of *Faust* to Gounod's contemporaries lay precisely in those qualities in which we now find it most lacking—in naturalness, simplicity, sincerity, and directness of emotional appeal. This can only really be understood if one compares *Faust* with either *Les Huguenots* or *Guillaume Tell*, both of them indisputably great works in their way but so overburdened with spectacular ballets, ingenious stagecraft, brilliant and pompous orchestration, and theatricality of emotion that it is easy to contrast the appeal—I will not say of Gretchen, for she can hardly be said to appear—but even of the rather insipid Marguerite. Despite her false flowers and her falser jewels she has a simple heart, and after the heroism and devilry of Meyerbeer and Rossini a simple heart was very engaging.

It was criticisms such as that published by Scudo on *Philémon et Baucis* in the *Revue des Deux Mondes* which deflected Gounod from his natural path and sent him in pursuit of the same aims as Meyerbeer. 'Les détails de la forme', wrote Scudo, 'les ciselures de l'instrumentation, les mièvreries du style ne suffisent point pour faire vivre une composition dramatique où la passion, les idées franches et la variété des couleurs ne brillent que par leur absence.' This was a direct challenge to give up his own careful, sober, discreet, and only faintly saccharine style for the tumult and the shouting, the captains and the kings of grand opera. Gounod's answer was *La Reine de Saba* and *Roméo et Juliette*, in which his own personality is overlaid, if not completely absent, and every possible concession is made to the Scudos of this world. But his capitulation was not final or complete. After a visit to Italy and Provence in 1863 he produced an enchanting *Mireille*, with music as fresh, winning and unsophisticated as its heroine—certainly one of his finest works, if not his finest, which still lives in France but seems quite unaccountably forgotten elsewhere. The scale is larger than that of the chamber operas, but the freshness of inspiration and (to use a contemporary term) melodiousness is, if anything, greater.

When the Franco-Prussian war broke out in the July of 1870 Gounod was a man of fifty-two. *Faust* (1859), after a not very successful start, had swept Europe and made him an international figure, giving him a reputation which *Roméo et Juliette* maintained even if it did not enhance. In France, with Berlioz dying in 1869 and Saint-Saëns still known just for his early orchestral works and that by a small minority only, his position was unchallenged. In September 1870 he fled across the channel to England, from which he did not finally return until 1874. In those four years he developed a new side of his personality but lost his position as leader of French music. His admiration for Mendelssohn and the whole attitude to music which he had learnt from Fanny Mendelssohn seemed to revive in England, where he found a taste for choral and religious music—and especially that of Mendelssohn—such as he had never met before. There was a demand for semi-religious drawing-room ballads which he could and did supply, and for oratorio to which he soon began to turn his attention. He produced an elegiac cantata, *Gallia*, inspired by the plight of France and, in 1873, a patriotic *Jeanne d'Arc*. But when he returned, he found that new forces had been mobilized in the French musical world while he had been absent and that his position, at any rate as leader of the younger generation, was gone. The foundation of the Société Nationale in 1871 was the beginning of that national renaissance in French music which soon began to bear such copious and remarkable fruit. Saint-Saëns was at its head and the original one hundred and fifty members included Bizet, Massenet, Lalo, Dubois, Bourgault-Ducoudray, Castillon, Franck, and Widor. These were the men who were to produce

the important works of the 1870s and 1880s, and Gounod's was
not the decisive influence in their musical development.

Of the three operas written before he devoted himself finally
to oratorio, *Cinq Mars* (1877) is already a step removed from
the purely operatic, *Polyeucte* (1878) a conscious attempt to
unite opera and oratorio, while *Le Tribut de Zamora* (1881) is a
completely unsuccessful attempt to recapture the old melodrama
beloved of the fifties. Gounod was in a state of transition. Saint-
Saëns has stated very well in his *Portraits et souvenirs* the
general qualities of Gounod's music and the particular ideals
towards which he was developing during the seventies:

The achievement of expressiveness was always Gounod's main preoccupation: that is why there are so few notes in his music . . . each note
sings. For the same reason instrumental music, pure music was never his
forte . . . His great desire was to discover a beautiful colour on the
orchestral palette and, in his search for this he refused to follow the ready-
made processes of the acknowledged masters, but carried on his experiments directly, studying the various timbres, inventing new combinations
and shades of colour to suit his brush. 'Sonority,' as he once said to me, 'is
still an unexplored country'.

These were always Gounod's chief traits, and the transition
period of the 1870s was really little more than a period of
groping towards a new way of realizing what were fundamentally
the same ideals.

He was anxious to reduce the number of modulations to a minimum, with
the idea that the composer should not make light use of such a powerful
means of expression. . . . [His ideal was] to obtain the maximum effect
with the minimum apparent effort, to reduce the representation of effects
to mere indications and to concentrate all the interest on the expression of
feeling. . . .

Admirable principles, deriving plainly from the enthusiasm
for Gluck inculcated by his old master Lesueur and perhaps
from his talks with Pauline Viardot; and yet what were the
results? In *La Rédemption* (1882) and *Mors et Vita* (1884),
where Gounod felt that he had at last achieved the style at which
he was aiming, the effect—the total effect, that is—is one of
bland and tender platitude. In the words of Blaze de Bury, 'Il
s'écoute phraser'. This strain of self-consciousness was not new.
While he was engaged on the composition of *Roméo et Juliette*

Gounod wrote to a friend: 'Au milieu de ce silence il me semble que j'entends me parler en dedans quelque chose de très grand, de très clair, de très simple et de très enfant à la fois.' This consciously 'childlike' simplicity is always suspect in an artist, for it almost invariably conceals an element of pomposity and insincerity which tends to grow with age, and even more with success, until it finally rots the sounder elements in an artistic nature. And in fact as he grew older Gounod did suffer from what might be described as the same *cher grand maître* complex as infected Hugo and Tennyson. Not content with being artists, these eminent Victorians were inclined to pose as prophets and, in proportion as their 'message'—the actual content of their works—became thinner, the manner in which they stated it became more and more sublime, more portentous, and more hollow-sounding. Gounod after 1870 might well have echoed Tennyson's despairing cry that he was the greatest master of English living and had nothing to say. This bromidic, self-consciously oracular state of mind is revealed in many of the stories which are told of Gounod's old age. 'Les enfants,' he is said to have remarked to a friend, 'ce sont les roses du jardin de la vie.' And to a woman who had accompanied him to the performance of some new work and asked his opinion: 'Qu'en pensez-vous, cher maître?' 'C'est rhomboïdal', which called forth the immediate and deserved reply: 'Ah! cher maître, j'allais le dire.'

In 1893 Gounod died; and although for the last thirty years of his life he had been outside the mainstream of French musical life, his death meant the disappearance of one of the great figures of nineteenth-century French music. For he was more than an individual composer: he was the voice of a deep and permanent strain in the French character. Actual pupils he had none; but a whole range of emotion, which had been voiceless before, had found in him its ideal expression, and his influence will perhaps never quite disappear for that reason. A few years before his own death Bizet complained that he was unable to hold his own against Gounod's influence; and in *Les Pêcheurs de Perles* and *La Jolie Fille de Perth* this influence is again and again noticeable. Even in *Carmen* it is not absent: Micaela is a pure Gounod character and her duet with José in Act I and her 'J'ai dit que rien ne m'épouvante' in Act III could never have been written if Gounod had not found her emotion a voice. In Massenet the strain of Gounod's influence is still strong, though so exclusively concentrated on the tender and the erotic that he was given the apt sobriquet of 'la fille de Gounod'. In his earliest works—the cantatas *Marie Magdeleine* (1873) and *Eve* (1875)—he expressed just that same half-mystical, half-erotic emotion which d'Indy was to flay as 'l'érotisme discret et quasi religieux'. *Hérodiade, Manon, Werther,* and *Thaïs* are the works of a man in whom an unbridled desire to please has conquered almost every other consideration and certainly removed Massenet from the fresher and more idealistic side of the Gounod tradition; yet in many turns of phrase, the charm and the *tendresse* of the melody there is no denying the influence of that tradition. It was this element of his great talent which Massenet transmitted to his many pupils, this rather than his own personal weakness; and so in the music of the many composers who came under Massenet's influence at the Conservatoire between 1878–96, Gounod's spirit was perpetuated. Romain Rolland in *La Foire sur la place* (*Jean-Christophe*, Vol. V) makes his young German musician complain of the moments in Debussy's *Pelléas et Mélisande* when 'the Massenet slumbering in the heart of every Frenchman awoke and waxed lyrical'; and it is a reasonable complaint. But Massenet would never have caught and perhaps immortalized that peculiarly French emotional mood which is summed up in the music of Des Grieux in *Manon*, for instance, if Gounod had not first given it a purer and less popular expression in *Faust* and *Mireille*.

If Massenet represents a degradation of the Gounod tradition, Fauré represents that tradition refined, ennobled and rarefied. Saint-Saëns noticed the bourgeois strain in Gounod and assessed its value and its danger:

. . . Son écriture, d'une élégance impeccable, couvre parfois un certain fonds de vulgarité . . . c'est comme un fonds de sang plebéien, mettant des muscles en contrepoids à l'élément nerveux dont la prédominance pourrait devenir un danger . . . c'est l'antidote de la mièvrerie.

In the music of Fauré it is not the 'élément nerveux' which predominates, but a kind of classical refinement, a sobriety and a quiet distinction, that 'fantaisie dans la sensibilité' which was the hallmark of the French musical genius in the eyes of

Debussy. Gounod's music was out of the direct line of the French musical tradition inasmuch as the primary appeal of at any rate his most popular works—*Faust, La Reine de Saba, Mireille, Roméo et Juliette* as opposed to *Le Médecin malgré lui* or *Philémon et Baucis*—was 'obviously emotional rather than intellectual and imaginative'. Fauré reversed this and by so doing garnered the harvest which Gounod had sown into the storehouse of the national art. We have seen Gounod finding inspiration in the German classics, musical and literary, in Mendelssohn and even, indirectly, in the German romantics. His individual style bears traces of all these foreign origins, French as it unmistakably is. Fauré, the most innately French of all French composers, assimilated this style, foreign elements and national alike, purified it of the emotional overemphasis which was due partly to Gounod's temperament and partly to the romantic era in which he grew up, and produced the classical French style of modern times. Yet Gounod's best songs and orchestral music were unearthed again after 1918 by 'Les Six' and proclaimed as valid French models.

Gounod himself was more than vaguely aware of the new forces at work beneath the surface of French music before his death—aware and by no means hostile. He begged for a reasoned and sober attitude—'ni Wagnerophobie ni Wagneromanie'—towards the music of Wagner, of whose greatness he was certainly aware, though he deplored the excesses to which his admirers and his detractors were equally prone to go.

La France est essentiellement le pays de la netteté, de la concision, du goût, c'est à dire l'opposé de l'excès, de l'enflure, de la disproportion, de la prolixité. La préoccupation—j'allais dire, la duperie—du transcendental peut, à force de persistance, arriver à nous donner le change, je veux dire à nous faire prendre le gros pour le grand, le pesant pour le solide, l'obscur pour le profond, le brouillard pour le sublime. . . .

Here, on the one hand, is the charter of Gabriel Fauré's music. On the other we find Gounod writing to Charles Bordes in 1893, the year of his own death, after attending the first 'Semaine Sainte de Saint-Gervais' at which the great polyphonic classics of the sixteenth and seventeenth centuries were performed:

Il est temps que le drapeau de l'art liturgique remplace dans nos églises celui de la cantilène profane et que la *fresque musicale* proscrive toutes les guimauves de la romance et toutes les sucreries de la piété qui ont trop longtemps gâté nos estomacs. Palestrina et Bach ont fait l'art musical, en sont pour nous les Pères de l'Eglise: il importe que nous restions leurs fils et je vous remercie de nous y aider.

There spoke the young enthusiast who, nearly fifty years before, had shocked the congregations and puzzled the good fathers of the Missions Etrangères in the Rue du Bac. Unlike so many 'great' men, Gounod remained to the end true to his first enthusiasms, and he saw in the new schools which were arising in his old age—even in the first revolutionary flutterings of Claude Debussy—legitimate developments of just those principles which had inspired him, principles which he could with justice claim to have reintroduced into the French musical consciousness.

8

FAURÉ AND THE HELLENISTIC
ELEMENT IN FRENCH MUSIC[*]

Classical antiquity has played a larger part in the literary imagination of France than of any other country. Indeed, the French have often considered themselves the natural heirs of the Greek philosophers and poets by virtue of the clarity of their ideas; the logical nature of their mental processes; and their instinctive recoil from excess of any kind in the aesthetic sphere. At the same time in their language, their institutions, and their military prowess, they have been very aware of their Roman antecedents.

In no other country but France could the seventeenth-century debate between the partisans of the 'ancients' and the 'moderns' have been so real or so impassioned; and the emergence of French classical tragedy in the middle of that century was to mean the all-but-exclusive attachment of French literature to the subject matter and the aesthetic principles of classical antiquity for something like one hundred and fifty years.

As if in revenge for this long subservience to the classical ideal, Romanticism in France was marked by its violently anti-classical character: the Gothic north or the Near East replaced the Mediterranean world, and the Middle Ages were rediscovered as a source of inspiration, completely ousting Greece and Rome. Shakespeare, whose 'disorderly' genius had so shocked the classicist Voltaire, was proclaimed as the patron saint of Romanticism; and translations of Goethe's *Faust* and even of E. T. A. Hoffmann's tales became fashionable.

As often, Berlioz provides an excellent guide to his country-men's true character, combining as he did a passion for the new 'romantic' conception of art with a persistent and profound attachment to the world of classical antiquity. In his pantheon,

[*] BBC 4 November 1973.

Virgil and Gluck took their places by the side of Shakespeare, Goethe, and Beethoven; and *Les Troyens* was the crowning work of his life.

This classical subsoil, as it were, in Berlioz's artistic nature was common to many Frenchmen. As early as 1853, classical antiquity was to appear in Leconte de Lisle's volume of *Poèmes antiques*; while in the same decade, Gounod wrote his first operas on classical subjects—*Sapho* and *Philémon et Baucis*—and Offenbach his *La Belle Hélène* and *Orphée aux Enfers*. This was a return not to the world of classical heroes and monarchs, but rather, in Gounod's case, to a romantically humanized vision of antiquity; or, in the case of Offenbach, to classical mythology as a mask beneath which political satire could be conducted.

It was in 1865 that Ernest Renan, on a visit to Athens, composed his *Prayer on the Acropolis*, a hymn to the classical ideal but contrasting the limitation of the classical virtues of reason, order, and moderation with the wider scope of the irrational worlds of poetry and religion. 'Nobility!' he exclaims. 'Simple and truthful beauty, whose cult is reason and wisdom, and whose temple is an undying lesson in conscience and sincerity . . . ! I will tear from my heart every fibre that is not reason and pure art. . . . I will cease loving my own sickness and delighting in my own fever. Support my firm resolve, O saving one! Help and sustain me!'

Renan's view of romanticism as a sickness, a fever all too delicious to those who caught it, of course echoes Goethe. And by the time that he had written those words, Paris had already experienced the music which, for the next half century, was to represent the antithesis of the classical ideal: the music of Richard Wagner.

It can safely be said that for thirty years after the first performance in Paris of *Tannhäuser*—that is to say from 1860–1890—any association with classical antiquity, such as in the choice of a song-text or the subject of an opera, was the hallmark of an anti-Wagnerian, a 'reactionary'. The best examples of this are Gounod's opera *Polyeucte*, based on Corneille's tragedy; and three of Saint-Saëns's symphonic poems, with their classical titles.

Early in the 1890s, the charms of the late Greek, or Hellenistic world of Alexandria were discovered. Anatole France's novel *Thaïs*, the story of an Alexandrian courtesan, appeared in 1890; and four years later, it was chosen by Massenet for what proved an enormously successful opera. Also attracted by this Hellenistic world, and particularly by its moral ambiguities, was Pierre Louÿs, a close friend of Debussy who set his *Chansons de Bilitis* in 1897. Debussy had in fact already paid his homage to this Alexandrian revival in his *Prélude à l'après-midi d'un faune*, based on a poem by Mallarmé; and Greek titles recur throughout his later works—'Danseuses de Delphes' and 'Canope' in the piano preludes, *Syrinx* and the *Six Epigraphes antiques*. Ravel's *Daphnis et Chloé* belongs to the same world; and so too do a number of Roussel's works—*Joueurs de flûte*, *La Naissance de la lyre* and at least eight of the songs.

Fauré's tribute to this fashion in taste was rather different. That he paid it at all is proof that he never belonged to the Wagnerian faction in French music. His contemporary, Saint-Saëns, wrote a successful classical opera, *Phryne*, in 1893; but Fauré's early settings of Leconte de Lisle's 'Lydia' and 'La Rose' pay hardly more than token tribute to the conscious simplicity and shapeliness of the classical ideal. It was not until much later, when his music was taking on that spareness of texture, economy of statement, and dry luminosity of harmony that we associate with both old age and the classical ideal that Fauré seems to have been strongly attracted by Greek subjects. His two operas both have Greek stories—Prometheus and Penelope—and some of the finest songs in the last three cycles are Greek in inspiration.

But what precisely do we mean when we discover in such music qualities that seem to recall those of Hellenistic art? For that art had lost the grandeur, the vital scope, and the intellectual power of Greek art in the fifth century BC: it was a lesser, more charming, more sensual art, without the religious or philosophical preoccupations of the earlier age but more humane, coloured with both scepticism and a world-weariness—essentially the art of an end-period of civilisation. Yet, for all this, it had preserved those older Greek characteristics of economy, clarity, and sobriety of expression. And it is precisely the combination of these with a sceptical humaneness that we find in the last works of Fauré. His *Pénélope* recreates not the world of Homer—primitive, instinctive, 'once-born'—but that world as seen through the eyes of a much later, reflective and sophisticated civilization, for which tragedy had lost its religious connotations and become the Virgilian 'lacrimae rerum'—the tears at the heart of things.

PART III THE WORLD OF MUSIC: ARTICLES
FROM THE *DAILY TELEGRAPH*

9

STYLES

Art Nouveau*

Music still has to borrow much of its language from the other arts, and in many cases what began as a metaphor has by now lost its metaphorical character; so that we speak for instance, without any sense of borrowing from the visual arts, of colour, line, and texture and no longer feel the literary implication when we speak of phrasing. In wider terms the analogies revealed by our borrowings are evidence of the direction in which music has developed. 'Baroque', for instance, was originally a term applied to architecture and only later applied to the other arts—first painting, then literature, and finally music. The great portmanteau word, 'romanticism', used indiscriminately of all the arts, was by derivation a literary term.

During the past few years, when the art of the 1890s has been revived, it has become increasingly common to find writers about music applying the term art nouveau (or, in Germany, *Jugendstil*) to music as well as to visual designs. Historically the art of the nineties represented the triumph of ideals first adumbrated in this country by the Pre-Raphaelite Brotherhood and given less provincial, more worldly, and much wider significance by the Aesthetic Movement in the eighties.

Art nouveau inherited from Pre-Raphaelitism an ideal of dreamy-eyed, slightly lymphatic femininity whose fleeting, 'willowy' (the word itself is revealing), mock-chaste forms in fact represented a new sophisticated sensuality, seen at its most unmistakable in the drawings of Aubrey Beardsley. The reaction against the obviousness, the crude realism, and frequent pretentiousness of late nineteenth-century art and life was seen at its clearest in the muted colours and flowing, vegetation-

* The *Daily Telegraph* 2 August 1969.

inspired lines of Morris's wallpapers and again in women's fashions, which revealed the figure naturally instead of hiding or monstrously exaggerating it with crinoline or bustle.

Where do these new aesthetic ideals first make their appearance in music? It is not too early, I believe, to see them in the music that Wagner wrote for his Rhinemaidens, figures which suggest immediately a relationship with the figures on the metal plaques that used to adorn the entrance to Paris metro stations and are classical examples of art nouveau. The repetition of a sinuous linear pattern and rocking rhythms was associated in the composer's mind with water and has obvious parallels in the plant world. Strauss's *Salome*, growing directly from Oscar Wilde's drama, is a plain instance of a Beardsley drawing elaborated and enlarged, sometimes out of recognition, by a musician whose sanguine temperament and brilliant technical accomplishments really allied him to an older generation of composers. Mahler's deliberate search for a quasi-popular simplicity in the *Lieder eines fahrenden Gesellen* and many symphonic movements (as in the tellingly named *Schlichte Weisen* of Max Reger) is a reflection of *Jugendstil* sympathies; and he revealed these unambiguously in his admiration for Charpentier's *Louise*, a classical example of the style in opera.

Debussy's earlier works are the most perfect instances of the art nouveau ideal in music. The *Deux Arabesques* and *Suite bergamasque* for piano have the characteristic watery, willowy gracefulness of line combined with simplicity of language. *Danse sacrée et danse profane* might have been inspired by the iridescent glass bas-reliefs produced by Lalique and so might the later *Six Epigraphes antiques*. In *Pelléas et Mélisande* we meet the same world of masked, implicit sensuality that can be traced back at least as far as the paintings of Burne-Jones.

Abolition of the will, the sense of human beings as puppets unable to understand or influence their own destiny, found perfect musical expression in Debussy's understatements, the short unemphatic phrases and the deliberately mysterious suggestions of the orchestral writing, in which muted tones and dynamics and tonally ambiguous ('whole-tone') harmonies play an important part. Maeterlinck captured the imagination of Schoenberg (*Pelleas und Melisande*) and Dukas (*Ariane et Barbe-Bleue*) as well as Debussy; and the same image of the human being as a helpless puppet at the mercy of unintelligible forces attracted Berg in Büchner's *Woyzeck* and Wedekind's *Lulu*, though Berg's mature musical language has few traces of the *Jugendstil* ideals which prompted such early works as *Sieben frühe Lieder* and the Piano Sonata.

This same passive, pessimistic atmosphere expressed in the arabesque lines and melting pastel colours of art nouveau pervades Delius's music. The opening of *A Village Romeo and Juliet*, for instance, is very close to the first of Debussy's *Deux Arabesques*; and the Dark Fiddler, an Arthur Rackham illustration, personifies the mystery against which the two children— more Pelléas and Mélisande than Romeo and Juliet—are helpless to defend themselves.

In all this music which can be generally described as being under the influence of art nouveau, rhythm is the weakest element. For rhythm is the life-blood of music and what this whole movement lacked was in fact vitality, or rather the vitality needed to inform anything but miniatures. It was a typical end-product of an exhausted civilization; and the renewal of music in the years following the 1914–18 war was largely the work of Stravinsky and Bartók, in whose music rhythm plays a very strong, if not dominant, part. Today this large-scale rhythmic design is once again almost unknown, and the fragmentation of the musical language has brought with it the confining of rhythmic structure to minimal proportions; while colour—no longer provided by harmony but by instrumental timbres—is again dominant.

Pastoralism[*]

The Commonwealth Festival of the Arts has brought performances of English music once popular and well thought of but not often heard recently in metropolitan programmes. Parry's *Blest Pair of Sirens*, in which I remember singing as a schoolboy, certainly possesses the noble, strenuous, masculine qualities admired by the Late Victorians. Although deeply indebted to both Mendelssohn and Brahms, this music was undoubtedly a real achievement in its day, and Parry's love of rhetorical sequences was inherited by Elgar. Not so his orchestration,

[*] The *Daily Telegraph* 25 September 1965.

mercifully, whose stolid unimaginativeness and unrelieved density reveals the amateur in this field and quite explains Parry's insistence on the 'black and white test' (reducing an orchestral score to a piano skeleton) as an infallible guide to a work's real quality. Among other works heard, Bliss's *Pastoral* for mezzo-soprano, choir, flute, timpani and strings, dating from 1928, proved of particular interest because it is one of the best examples of a genre which enjoyed an astonishing popularity with English composers between the wars.

On the face of it, the pastoral seems the most improbable of all fields for the artists of a highly industrialized society to explore. The genre has always been something of an affectation— highly civilized, urban poets like Theocritus and Virgil in antiquity, or Tasso at the Renaissance, and French court poets and composers during the seventeenth and eighteenth centuries, singing the praises of the simple farmer's life or the so-called innocent and improbably refined amours of shepherds and shepherdesses. In fact the whole 'pastoral' convention in the arts has always been one designed by, and for, those who for one reason or another preferred convention to reality. It has been the vehicle of much charm, ingenuity, and refined sensuality; but it is significant that among composers in the past, it has attracted only very lesser men. Mozart's youthful *Bastien und Bastienne* was only a parody of Jean-Jacques Rousseau's enormously successful *Le Devin du village*.

The discovery of folk-songs, which had such a deep effect on Vaughan Williams's music and so set a whole fashion, is not in itself enough to explain the recrudescence of this sentimental ruralism in the twentieth century. The Bohemian and Russian composers whose art was fertilized by the folk music of their countries, wrote no pastorals; and there was certainly nothing nostalgic or escapist in the use to which Bartók put his folk-song studies.

English composers after the 1914–18 war turned Nelson's blind eye to what was happening to music on the Continent. The Channel still existed in those days and it was seriously believed that what interested musicians in Prague, Vienna, Budapest, Paris, or Berlin had not necessarily any relevance in London. Ignoring the new music, English composers in most cases chose also to ignore modern life. The tourists' England of thatched cottages, Beefeaters, Changing of the Guard, and royal and legal pageantry had its intellectual and aesthetic counterparts in the neo-medievalism of Chesterton and Belloc, and Vaughan Williams's *Hugh the Drover, Fantasia on Christmas Carols,* and *Five Tudor Portraits*; in Moeran's rhapsodies, Walton's *Crown Imperial* and *Portsmouth Point,* Ireland's *Mai-Dun, Concertino Pastorale,* and *London Overture,* Bax's Celtic fantasies, and Warlock's skilful Elizabethan pastiche. The music of Delius, himself a very conscious 'yea-sayer' to life in general, won great favour and influence with a generation which was in fact passionately rejecting life and music in their modern forms and turning instead either to a sentimentally coloured past or to a self-indulgent contemplation of the pathos of human mutability.

It would not escape the notice of a Marxist observer—and was undoubtedly a strong contributory cause of this situation—that all these composers belonged either by birth or adoption to the upper middle class, whose comforts and glories did indeed lie in the past while the present was comparatively bleak and the future bleaker still. Sir Hubert Parry himself, a free-thinking Etonian of an earlier generation, had inaugurated the strange tradition by which unbelieving English composers have voluntarily chosen to set religious, and even liturgical texts, presumably a form of nostalgia for a faith emotionally desired though intellectually rejected.

Such factors were an undoubted source of weakness in this generation of composers, if only because emotional and intellectual compromise debilitates all artistic creation. It is significant that the English composers who have competed on an equal footing with their Continental colleagues, and been accepted, are precisely those in whom this very English spirit of compromise is least noticeable.

Pastiche and Eclecticism[*]

There is a cautionary tale, current among critics, of a virtuoso being shown a notice in which his playing was described as 'incandescent'. Like many outstanding performers, this virtuoso was no great master of words and the nice distinctions between

[*] The *Daily Telegraph* 7 January 1967.

their meanings: 'incandescent' conveyed nothing to him. 'Is it good?' he asked his agent.

The vocabulary of words which can be legitimately applied to music or to performance is indeed limited—though not so narrowly as Beecham's 'twenty epithets'—and every effort should be made to increase rather than to restrict it. But some definition of terms will often be necessary.

In a recent interview, Malcolm Williamson reacted with predictable sharpness to the suggestion that his music is, or at least has been, 'eclectic'—a word not uncommon in criticism of the arts but probably conveying no more to many readers than (in virtuoso's basic English) bad. Malcolm Williamson suggested that an eclectic work was one in which 'the composer's own musical personality is less in evidence than the manifold influences which there are on all of us'. This may serve as a rough definition, useful because it emphasizes the fact that no art is, so to speak, chemically pure in the sense of being absolutely original.

Of course by far the greater part of all music committed to paper is, in this sense, so eclectic that it never reaches the public. Music publishers are inundated with scores that are pale reflections of fashionable models, and it is often easier to get an artistically worthless 'originality' performed (though not perhaps published) than a painstaking example of skilled craftsmanship that reveals no marked personality in the composer. A work of art is valued today just in so far as it is a personal statement, and the only permissible form of eclecticism is that which is deliberate. It was an admirer who imagined Poulenc saying to himself as he planned a new work:

I will have some Ravel here and some Chabrier there. Here I shall have a touch of early Debussy . . . then go back to my own Ronsard settings. This shall start like *Mavra*, that will be very Marie Laurencin.

Yet no one could ever suppose that the work composed on these jackdaw principles was by anybody but Poulenc; for no other composer would chose the same heterogeneous material or use it in the manner which we immediately recognize as Poulenc's. Here the material is eclectic but the finished work original, an indisputably personal statement. Stravinsky's borrowings of style or even material (in *Pulcinella* or *Le Baiser de la Fée*) never prevented him from producing completely individual works belonging unmistakably to the age in which they were written.

When they first appeared during the 1920s, Stravinsky's neo-classical works (to give them a rough but convenient classification) were often called pastiche. This was a complete misnomer. Pastiche is the imitation of the style of another composer or another age, such as Tchaikovsky attempted very successfully in *Eugene Onegin* (Triquet's couplets) and *The Queen of Spades* (the miniature dramatic cantata of 'The Faithful Shepherdess' in Act II, among other things). Stravinsky's and Poulenc's approach to other men's music was entirely different; and they would have been as annoyed as Tchaikovsky would have been pleased if their authorship had been questioned.

Having seen what eclecticism in music is not, can we discover a more positive example of what in fact it is? Hollywood has in the past provided what is perhaps the clearest, and is certainly the most familiar, instance; and one that constitutes what seemed at one time an irreparable damage to the reputations of two great composers. The indiscriminate pillaging of Rachmaninov and Ravel by the lesser screen-music-writers of the 1940s and fifties brought the works of both men into disrepute with musicians more sensitive than rational in their reactions, and ready instinctively to blame the original proprietors for the use that thieves were making of their stolen property. It was only when Hollywood fashions changed that the younger generation of music-lovers could approach Ravel and Rachmaninov without prejudice.

The Germans have adopted from Greek mythology the useful word *epigones*, unfortunately not properly naturalized here yet, for the lesser men who instinctively follow a great creator, often unconsciously adopting his language and even his mannerisms. In Germany, the Wagnerian *epigones* filled a whole generation of whom Humperdinck, the composer of *Hänsel und Gretel*, is probably the best known and certainly one of the most interesting. Stravinsky has been followed by a similar host of musicians composing, as it were, in his wake, while in this country first Vaughan Williams and then Britten have been accorded this doubtful honour.

It is a strange fact that 'Wagnerian *epigone*' has been a term of

abuse applied at a first hearing to several composers whose
works, when better known, prove to be something very different
from mere pale imitations of a greater model. Though it now
seems incredible, Bruckner was for a long term dismissed thus in
this country; and Elgar, as far as I know, still enjoys the same
fate in Germany and France. To be an *epigone* is to belong by
definition to a second generation; but, as we know from
revolutionaries and from those who change their nationality,
members of a second generation are often very far from
repeating the character—though they may reproduce the superficial characteristics—of the first.

Neo-classicism[*]

When did a simple ascending scale cease to be a musical event,
an 'aspiration', and become a formula, pigeon-holed automatically by the listener as a 'scale' and emptied of its emotional
significance? Or, in general terms, how does a language become
devalued, like a currency?

The great task of composers during the first half of the
twentieth century has been to discover some way to recapturing
the eventfulness of music simply as sound, of revivifying the
individual unit and the interval so that these become once again
events in the listener's consciousness. How complex a task this
had become by the beginning of the present century can be seen
by comparing the effect of, say, the opening two bars of Bach's
Italian Concerto—an octave leap down followed by a simple
ascending scale—or the unison scale passages that lead up to the
trumpet call in Beethoven's *Leonore No. 3*, with the enormous
paraphernalia needed by Ravel to invest with importance the
simple rising scale passages that form the climax of 'Ondine' or
the solemnity with which Sibelius surrounds a similar ascending
scale at the opening of his Seventh Symphony.

Those who accepted the fact that further harmonic complications could not disguise the exhaustion, the debasing of the
traditional musical language or, as it were, restore a virginity so
thoroughly debauched by the nineteenth century, were faced
with two alternatives. They could either turn to the past and to
non-European models to provide them with a fresh, though not

* The *Daily Telegraph* 29 June 1963.

in fact new musical material; or they could refashion existing material to form, in effect, a new language after the model of Esperanto and the other attempts at an international language. Both procedures were in fact arbitrary; but whereas the former followed many respectable precedents in the arts, the latter bore a closer resemblance to the scientific thought-structures which have been peculiar to this century.

Fifty years after this crisis we find, as we should expect, a *rapprochement* between the representatives of these two— originally sharply opposed—schools of thought. Stravinsky, who led the so-called 'neo-classical' party that hoped to revivify music by returning to the past or drawing on non-European models, has come to terms with Schoenberg's serialism and with Webern's emancipation of the single note—and rest. But Stravinsky's serial neo-medievalism is a personal solution that hardly satisfies the large public, though it has not been without its influence on young composers.

Serious music has certainly ceased to be commonplace, in the sense that neither the sounds themselves nor their sequence can be taken for granted. But in safeguarding against this danger composers have too often fallen into the opposite extreme. If the old language had lost its force, the new is too often unintelligible to the ordinary listener, who easily loses patience with any sequence of sounds whose logical coherence he cannot perceive. The great gain of the past fifty years has been the huge extension of sensibility to include ancient and exotic as well as genuinely new sound-combinations, a new alertness of ear and mind. Better that music should be puzzling and alarming than sweet and consoling, a lively young delinquent rather than a corpse in the trappings of a Californian funeral parlour. Better still, of course, that music should be both mature and lively, unexpected yet audibly coherent. Is it to our traditional national gift that Benjamin Britten owes his genius for this particular form of compromise, which was well described recently as 'the genius for expressing the extraordinary through ordinary means used with a difference'?

It seems as though the time had come for the consolidation of the past fifty years' gains, a process of sifting and hybridization (all mature art is eclectic in origin), and we can observe this not only in Britten himself but in the most promising of the younger composers, notably less doctrinaire than their seniors who lived through the years of active revolution. Doubtless there is still much experimenting to be done, but the time when experiment and creation were equated is over. The enlarged musical language of the mid-twentieth century, though still far from stable, is an instrument that invites the creation of the new 'mainstream' works that the public awaits, and composers are beginning to dare to write.

Baroque[*]

Anniversaries are popular; but one that is not very likely to be celebrated is the half-centenary of the term 'baroque' as applied to the music of the seventeenth and the first half of the eighteenth centuries. Curt Sachs, who introduced the term in Germany, following the art-historian Wölfflin, published his *Barockmusik* in 1919; and the unusually close tie between the visual arts and music during these one hundred and fifty years proved a strong recommendation for retaining its use.

Whatever its derivation—and the most probable seems to be from the Portuguese *barroco*, meaning an irregularly or fantastically shaped pearl—the French use *baroque* to mean queer, whimsical, uncouth, while to the Italians *barocco* means awkward, in bad taste, over-ornamented, in each case reflecting the judgement of a subsequent age on the art conveniently lumped under this heading.

But what, in truth, are the features of baroque art—visual, literary, and finally musical—which distinguish it from the Renaissance classicism from which it developed and the eighteenth-century 'neo-Palladian' classicism which, after the rococo interlude, replaced it? Eugenio d'Ors's theory is that 'baroque' traits can be found in all periods when a civilization is regressive or on the defensive. In this sense, flamboyant Gothic architecture, Wagner's *Tristan* (baroque romanticism) and even Strauss's *Der Rosenkavalier* (*Jugendstil* with baroque accessories) are all examples of a similar complex psychological attitude, for which parallels can also be found in Hellenistic, late Roman, and even Inca art.

The essence of the baroque is movement, unrest as against the

[*] The *Daily Telegraph* 21 June 1969.

static order of the classical ideal. The baroque artist is concerned not to create a perfect form but to move or dazzle his audience; and a good case has been made out for regarding the baroque art of the period that concerns us as a characteristic production of the Counter-Reformation in general and of the Jesuits in particular, whose Roman church of the Gesù is one of the earliest (1568) monuments of baroque architecture. Much of baroque art has this propagandist, rhetorical quality, showing the characteristics that distinguish not only the preaching and political literature of the day but propaganda in every age. Everything is consciously planned for effectiveness, and exaggeration is an intrinsic feature. Pathos replaces the ethos of classical art, whose psychic continuity gives way to change and excess—sensual, heroic, or devotional. Rhetoric replaces statement, grandiose display the graspable dimensions and emotional reserve of classical art.

Church music of the baroque period, like church architecture, was designed to transport the faithful from the world of everyday reality into a supernatural realm, where even the natural laws of physics appear to be superseded by visual and acoustic illusions, *trompe-l'oeil* and ingenious polychoral dispositions achieving similar 'supernatural' results.

But the most characteristic single creation of the baroque period is the opera, not simply as a musical form but as a colossal spectacle designed to move and impress an audience with the splendour and power of monarchy. More profoundly, opera corresponds by its very nature to the baroque ideal in that it is heteronymous music, not subject to its own laws but deriving these from the drama, whose emotions it must 'express' (baroque 'Expressionism' is a perfectly legitimate concept) or evoke, often in ways not fundamentally unlike those of later Impressionism. Opera, in fact, retained to the end of the nineteenth century and beyond many of the characteristics of a baroque work of art; and Pierre Boulez is not alone in thinking that opera as such is so essentially a baroque art-form that it has no further place in our musical life.

On the other hand, baroque art in all its forms, and the opera in particular, have during the past fifty years aroused a new and ever-increasing interest not only among scholars but among the general public. Purcell and Vivaldi have been succeeded by

Monteverdi and Schütz as darlings of the connoisseurs; and the line is plainly continuing with Cavalli and Telemann, whose rococo features are hardly more than a final form of the baroque, reflecting the ultimate triumph of ornament over design. This instinctive sympathy with baroque music in its most characteristic forms suggests that Eugenio d'Ors may well be right in associating all forms of baroque with civilizations that are contracting or on the defensive. Certainly in our own society the taste for baroque music is characteristic of a conservative, backward-looking mentality and is often consciously cultivated as an alternative to avant-garde music, as a 'new' music that is not newly-made but newly-discovered.

Much avant-garde music itself, except where it consciously or unconsciously assumes the form of non-music, has its roots in early twentieth-century Expressionism which, as we have seen, was a rediscovery of one of the root principles of baroque art. It, too, very clearly proclaims—may even be regarded as propaganda for—a message, though it is a negative one: the final death of that old Judaeo-Hellenic civilization, an earlier stage in whose battle for survival first prompted the rhetorical, propagandist art of the Counter-Reformation and extended almost to the heart of the eighteenth century as the 'Baroque Age' par excellence.

10

OPERA

The Fiery Angel[*]

In the eyes of his fellow-countrymen, Prokofiev redeemed his fifteen years of self-imposed exile in Western Europe and the United States (1917–32) by twenty-one years of devoted work as a consciously Soviet composer between his return to Russia and his death in 1953. In perspective, the break in his music between the two periods, Western and Soviet, is already beginning to seem less marked than it did to his immediate contemporaries. Thus the Sixth Symphony (1947) is seen to contain many of the 'Western' elements which the composer was said to have renounced on his return to Russia, while *Tales of an Old Grandmother* (1918) have many of the characteristics often believed to have been developed by Prokofiev only after his return to Russia.

Some of the productions of his years in the West, however, are not likely to find a welcome in his own country in any foreseeable future, and among them is his opera *The Fiery Angel*, which is being given its first stage performance in this country next week, by the Frankfurt Municipal Opera at Edinburgh. This was written between 1920 and 1927, for the most part at Ettal in Bavaria, where Prokofiev lived from March 1922 until the autumn of 1923, though he did not begin the final revision and orchestration until the summer of 1926. Bruno Walter was at one time interested in the work for Berlin, but Prokofiev soon despaired of a performance and used many of the opera's themes in his Third Symphony which Monteux conducted in Paris in 1929.

The book of *The Fiery Angel* is taken from a story by Valery Bryusov that appeared in the Russian Symbolist magazine *Vesy*

[*] The *Daily Telegraph* 22 August 1970.

('The Scales') in 1907–8. Bryusov was an important figure in the Russian Symbolist movement, which borrowed all that was most likely to outrage bourgeois sensibilities from the works of Baudelaire and Huysmans—the conception of woman as a poisonous evil and love as an object of execration as well as a source of ecstasy, interest in the macabre, the masochistic and all that was 'contra-natural'. In one member of the group, Sologub, this attitude developed into a conscious apotheosis of evil and perversion. There was much talk of 'the art of the lie' and Bryusov wrote to his friend Balmont: 'What I love in you is that your whole life is a lie.' It is not necessary to be an orthodox Marxist critic to speak of this group and their art as decadent.

The story of *The Fiery Angel* is presented in the third person and in the style of the German sixteenth century. The title runs:

A truthful narrative in which the story is told of a devil who more than once appears to a young maiden in the form of a fiery spirit and tempts her to commit various sinful acts; of godless dealings in magic, astronomy, and necromancy; of the trial of the young maiden conducted by His Grace the Archbishop of Trier; and also of meetings and conversations between a knight and the triune doctor Agrippa von Nettesheim.

The combination of diabolism and eroticism with hysteria and hallucinations had a curious attraction for Prokofiev, in whom there seems to have been at least a streak of that taste for the *faisandé* or 'high' (game, not drugs) that marked Diaghilev and the artists of the *Mir Iskusstva* group to whom Prokofiev appeared such a country cousin. Although he had played the piano as a boy, in the winter of 1906–7, at Walter Nouvel's Evenings of Contemporary Music in St Petersburg, it was not until ten years later that Diaghilev tried to sophisticate him by introducing him to avant-garde circles in Milan. Diaghilev believed that he failed; but Prokofiev's choice of a Bryusov story for an opera suggests that Diaghilev's failure was not perhaps so complete as he imagined.

There is some mystery about the first origins of the music of *The Fiery Angel*. When Prokofiev protested with some heat against the description of his Third Symphony as a programmatic work, he acknowledged that, although its principal themes were indeed derived from the still unperformed opera, they had originally been conceived instrumentally; and that, in the Symphony, they were merely 'restored to the purely instrumental realm to which they belonged by right'. Certainly much of the hysterical Renata's music is in a muttered, even mumbled parlando style which leaves the strictly thematic interest to the orchestra. The first appearance of her love theme in Act I, for instance, is in the orchestra while she herself is engaged in a breathless parlando narrative. On the other hand, the Fortune-teller in Act I is almost traditionally Russian, and the actual prosody of the vocal parts are so closely connected with Russian speech-rhythms that translation is difficult and cannot sound natural.

The poltergeists in Act II knock on the walls against the background of a hectic orchestral toccata recognizably akin to music in *The Love for Three Oranges*. This fundamentally instrumental ostinato technique dominates the whole of the big Inquisitor's scene in Act V, where Prokofiev indulges to the full that sense of the macabre, the unreal, and the 'contra-natural' which attracted him in Bryusov's tale. These rhythmic ostinatos, which recur on a smaller scale throughout the work, are what chiefly distinguish the music from that of Hindemith's *Cardillac*, a work not wholly dissimilar in character and composed at almost exactly the same time as *The Fiery Angel*. Prokofiev and Hindemith both use harmonies predominately based on the interval of the fourth; their relationship to tonality is equally close in reality though often distant in appearance; and their writing for the voice shows a very similar reaction against the vocal styles of the nineteenth century, a deliberately awkward line marked by intervals that are vocally unrewarding even when they present no technical difficulty.

It is no coincidence that *The Fiery Angel* has been chosen in the West to represent Prokofiev's operatic output, rather than either *The Duenna* or *War and Peace*. An opera in which there is a scene of communal 'possession', or hysteria, in a convent and Mephistopheles eats a small boy and resurrects him from a dustbin could hardly fail to attract the favourable attention of an opera-house today—and at the same time to confirm, alas!, the official Soviet view of bourgeois society.

Fidelio[*]

German opera, which was to have such a splendid flowering, had a delayed and difficult birth, and its two first masterpieces remain to this day anomalies. Is Mozart's *The Magic Flute* a rubbishy pantomime for which the dying composer wrote, in something like a fit of abstraction, the supreme music that no commission, however unworthy, could prevent him writing? Or did the collaboration with Schikaneder unconsciously prompt Mozart to the last and greatest of those double-natured works whose appearance is childish but whose essence we may justly call angelic—if only because the child's play embodies a message as unmistakable as it is inseparable from the superficially trivial vehicle containing it?

And what is *Fidelio*? Analytically speaking, it consists of an act of French *opéra comique* (heavily loaded sociologically and politically, as this form had been since 1760), a central scene of high melodrama, and a final cantata or Hymn to Conjugal Love, such as French composers wrote under the Directoire. We know that Beethoven's model was in fact Cherubini, whose *Les Deux Journées* (1800) was a thoroughly up-to-date story of republican virtue and conjugal devotion that came oddly on the heels of his old-fashioned Gluckian tragedy *Médée*. But *Fidelio* is unique in the operatic repertory, where it is secure simply by virtue of the moral passion with which Beethoven suffused his score.

Perhaps only a bachelor, for whom marriage remained an unattainably ideal state (though he seems to have had some difficulties in not attaining it), could have conceived and maintained the high fervour of Leonore's music. Certainly Florestan consoling himself on the brink of death with the thought that he has done his duty was a more real figure in the era of Kant's Categorical Imperative than it is today, when the realities of political imprisonment may have become more horrible but are in any case better known.

The character of Pizarro presents another difficulty to an age all too familiar, often at first hand, with prison commandants. We know that the blustering, obvious villains, who roll their

[*] The *Daily Telegraph* 10 October 1964.

eyes and stamp their feet, are less horrible than the smooth-faced and cool-mannered who make a show of benevolence. If Beethoven and his librettist had given Pizarro even one redeeming feature—perhaps the famous affection for children, or an interest in the prison garden—we should find him more credible. As it is, it is almost impossible not to feel a twinge of pity for a man who, if we can persuade ourselves to believe in him at all, was so plainly mentally sick and so easily defeated.

In fact, although the essence of the *Fidelio* story still has power to move us, it is only the passionate sincerity of Beethoven's music that will make us accept the psychological *naïveté* of its details. This *naïveté* was, as we can see clearly from his letters, a very important trait in Beethoven's character, and one that played a large part in prompting the reaction against his music forty years ago, when Mozart's stock began to rise so significantly. Beethoven's music in general, and *Fidelio* in particular, makes a unique demand on performers. A flawless technique, a beautiful tone, and unfailing musical instinct are simply not enough without a capacity for spiritual enthusiasm, a kind of moral inflammability that is by no means universal among artists. Would any connoisseur choose Heifetz as the ideal soloist in Beethoven's Violin Concerto? And have not all great performances of *Fidelio* depended on precisely this quality in the principals? In fact its absence destroys the character of the work, showing up Beethoven's dramatic miscalculations without providing anything to counterbalance them.

A single example: the canonic quartet 'Mir ist so wunderbar' in Act I. Objectively considered, this is too serious and profound a reflection for three of the four characters who make it, and the listener will be instinctively aware of this unless he is caught up in a moment of rapture that can only be communicated by singers who themselves respond to the serene moral penetration of Beethoven's music.

The comparative clumsiness and *naïveté* of Beethoven's presentation of the Leonore–Florestan story in terms of the thèatre can be seen if we compare the opera *Fidelio* with the *Leonore* tone-poems. In these Beethoven presented in purely musical, as it were abstract, terms the same story that he tells in concrete, theatrical detail in *Fidelio*. (That is why it is a painfully inartistic tautology to perform *Leonore No. 3* between the two final scenes.)

Beethoven, whose music is dramatic by its very nature, was nevertheless always irked by the sheer mechanics and limitations of the theatre. Like Berlioz after him, he could penetrate to the heart of a dramatic situation in purely musical terms and was embarrassed by having to use words and physical actions. It is not for nothing that Leonore and Florestan speak of their joy as being inexpressible in words ('O namenlose Freude'), for it could in fact be expressed only in music. And the effect of Florestan's aria when it appears wordless in *Leonore* No. 3 is far deeper and more lasting than when it is heard in the theatre, wedded to Sonnleithner's oddly stiff and priggish-sounding words and sung by a tenor chained to the floor of the stage. Beethoven's music transcends the theatre of bricks and mortar, and it is for that reason that *Fidelio* makes unique demands on its performers.

Simone Boccanegra*

A good case has been made for considering Verdi the last of the great composers to belong to the 'naïve' class, in Schiller's sense, which distinguishes artists into 'naïve' and 'sentimental'. This would not, however, lead us to suppose that Verdi was naïve in the commonly accepted sense; for this son of peasant parents had his full share of the hard-headed shrewdness which shows itself in bargaining—whether for a cow or a libretto—and in an unwillingness to commit oneself finally to any course of action—whether buying a piece of land or engaging a singer—without first obtaining firm evidence of quality.

It may well have been this trait, instinctive rather than conscious, that prompted Verdi in 1880 to turn to Arrigo Boito when he wished to revise his *Simone Boccanegra* which had failed twenty-two years earlier in Venice. Since June 1879, Ricordi had been urging, and Boito himself preparing, the libretto for that collaboration which was finally to issue in *Otello*. Verdi may well have thought that Boito's ideas on the salvaging of *Simone Boccanegra* would provide a good specimen

---

* The *Daily Telegraph* 26 April 1969.

of his understanding of theatrical problems; and that a collaboration in this minor undertaking would serve as the preliminary canter which would decide whether or not Verdi should buy the horse.

Piave's original libretto for *Simone Boccanegra* was as naïve, as improbable, and as difficult to follow as his *Il Trovatore*, with which it had many features in common, both being founded on stories by the Spaniard Gutiérrez, who specialized in stolen children and mistaken or disguised identities. Boito described it in a letter to Verdi as 'a rickety table' with one leg needing adjustment; but he soon came to see that as soon as one leg was adjusted, another put the table out of balance. Verdi refused Boito's suggestion of a whole new act, as involving too much work; but in the new council-chamber scene at the end of Act I, he not only wrote some of his finest music but went some way to repairing a major blemish of the drama. For here he showed the *condottiere* Boccanegra in his true sphere, that of action, instead of either lamenting the past or exercising improbable powers of forgiving his enemies—neither of them characteristic of such a man.

We are still left with the awkward equations Amelia Grimaldi = Maria Boccanegra (and the very thin reasons for the alias) and Andrea = Fiesco (and the much greater problem of who among the persons of the drama are aware of this identity). We may find it hard, too, to remember that Amelia is Fiesco's granddaughter as well as Boccanegra's daughter; and the interval of twenty–five years between the Prologue and Act I is difficult, if not impossible, to imprint, visually, on the audience's imagination.

Almost the same interval separated Verdi's original music from the new scenes which he wrote for Boito's revision. The musical similarities to *Il Trovatore* are strong in the Prologue, where Paolo's 'L'atra magion vedete' recalls Ferrando's narration and the final chorus sounds oddly thin and conventional; and in Act I, where Gabriele's offstage aria recalls Manrico's first appearance, Amelia's breathless 'Ei vien! l'amor m'avvampa' is very close to Leonora's 'E deggio e posso crederlo' in the finale of Act II and 'Ripara i tuoi pensieri' resembles in key and rhythm as well as mood 'Ai nostri monti' in Act IV. After the recognition duet (Amelia and Boccanegra) the primitive rhythm and hissed whispers of the conspiratorial duet between Paolo and Pietro make an almost comic effect, following the broad, expansive melody of 'Figlia! a tal nome palpito' which seems to foretell Desdemona's 'Io prego il cielo per te' in Act II of *Otello*. The new scene in the council chamber, which shows the interplay of public interests and private passions, the grand gestures of heraldic fanfares and crowd acclamations interwoven with lyrical passages and venomous asides, similarly suggests comparison with the finale of Act IV in *Otello*.

Boccanegra's great appeal ('Plebe! patrizi!') is couched in the two extreme keys (E flat minor and G flat/F sharp major) which play an important part in Verdi's new score and again suggest comparison with *Otello*—the opening of the love duet (G flat major) and the A flat minor heart of the tragedy at 'Dio! mi potevi scagliar'. Boccanegra's plea for peace between Genoa's rival parties has a beauty of line and an intensity of expression that rival Rodrigo's farewell to Carlos; and Amelia's repeated cries of 'Pace!' floating above the chorus belong to the same transfigured, F sharp major world as Elizabeth's 'S'ancor si piange in cielo' in Act V of *Don Carlos*. By a masterly musico-dramatic stroke, the single line heard in the orchestra as Boccanegra indicts Paolo is a diminished mockery of this expansive, ethereal phrase of peace. Amelia uses the same key, and similarly wide-spaced phrases in her plea for Gabriele's confidence in Act II ('Sgombra dall' alma il dubbio'), just as Boccanegra's E flat minor is sympathetically imitated by Gabriele ('Perdon, perdon, Amelia') and by Fiesco ('Delle faci festanti' and 'Piango'). In each case the situation is a sudden reconciliation, as though Verdi associated this key with the concept of peace obtained by an admission of guilt.

The magnificent often sinister unisons which Verdi in this late version associates with Boccanegra form a striking feature of the score. They range in character from the fulminating *tutta forza* introducing the curse of Paolo to the soft, almost paternal solemnity of the opening phrases of the second act's Finale and the long passage—carefully characterized by the composer's many marks of expression—which heralds the dying Boccanegra's final entrance.

Simone Boccanegra, mostly dark in colouring and grave if not tragic in character, is relieved by Verdi's great affection for the natural beauties of Genoa and its coastline. The orchestral picture with which Act I opens is all sunlight, air, and freedom and might be regarded as the counterpart to the opening of the moonlit Nile Scene in *Aïda*, both founded perhaps on memories of Meyerbeer's *L'Africaine*. And the dying Boccanegra's passionate apostrophe of the sea ('Che refrigerio! . . . la marina brezza!') may well be an echo of what Verdi himself had often felt when escaping from urban Milan or the inland flatness of Busseto and its surroundings.

Les Troyens*

The Royal Opera House has subscribed handsomely to the Berlioz centenary by reviving *Les Troyens*; and the response of the public suggests that the grandiose, the heroic, and the unfamiliar still combine to form a powerful attraction. It may of course be Covent Garden's familiar emphasis on the spectacular that has given this revival its somewhat improbable success. In any case, it is an interesting reflection on the opera-house's attitude to the work that the original revival was entrusted to Sir John Gielgud and the present one to Minos Volanakis, both men of the theatre with no previous experience of opera; and that the present sets and costumes were designed by Nicholas Georgiadis, who was responsible for the overpoweringly 'luxy'—there is really no other word for it—production of *Aïda*. The unspoken, implicit judgement which determines these choices is that *Les Troyens* is Parisian grand opera—with a difference perhaps, but not a difference that transcends categories. 'Theatre' and spectacle are given a very important role, and the many divertissements are put on with a lavish splendour.

This approach may be dramatically speaking legitimate; but it does no service to Berlioz since it emphasizes all that is weak and conventional in *Les Troyens*. Original in his handling of the orchestra, in his approach to melody, harmony, and musical form, Berlioz found it impossible to escape the operatic conventions of his day except in details which may modify, but do not really alter, the conventionality of the whole; and so it is isolated moments of the score that are memorable rather than the work as a whole. The proportion of purely exterior,

* The *Daily Telegraph* 18 October 1969.

inorganic scenes is very high—almost the whole of the Trojan part, except for Cassandra's scenes; most of the first Carthaginian act, the Royal Hunt, and much of the next act. We no longer find it easy to accept a large-scale opera in which there is so much decoration and generalization, so little real delineation or development of character.

It is because the great majority of the work is music of situation, observed from outside rather than lived from within, that (with the possible exception of Cassandra) Dido is the only character in *Les Troyens* who moves us, and then less in her formal response to Aeneas's passion than in her widow's scruples of disloyalty to the past. These prompt Berlioz to abandon for once his grand literary generalizations—for that is what the love-music really amounts to, for all its intoxicating sensuous appeal—and to individualize his heroine, to explore for a moment the woman under the regal mask.

There is a sense in which any production of *Les Troyens* is the theatrical performance of an oratorio, the exact opposite of the concert performance of an opera. For everything in the score that demands dramatic representation is really secondary, conceived in terms of French grand opera for which Berlioz lacked the necessary qualities—the readiness to subordinate musical to theatrical considerations and a sense of what is really effective in the theatre. Cassandra's tragic scenes, Dido's scruples and remorse, the classical tableaux of the love scenes; and such isolated, inorganic moments as the song of Iopas and Hylas and the *opéra comique* vignette of the two sentries belong to another world from the Alma Tadema 'classicism' of the crowd scenes and the ballet *divertissements*. There must have been other people beside myself who heartily echoed Dido's comment on these entertainments—'assez, ma soeur, je supporte à peine cette fête importune'. Indeed, Mr Volanakis's only unforgivable offence is to turn the 'Royal Hunt and Storm' into just another of these fêtes by using this most airy and imaginative of Berlioz's orchestral visions as the accompaniment for a disastrously pedestrian pantomime-style ballet of hideously garbed woodland spirits.

Here we are very near the heart of Berlioz's persistent failure as an operatic composer. This is rooted in the visionary, imaginative quality of his musical ideas, which come to complete life in the specific orchestral form in which he conceived them and are not capable of being further enhanced. Berlioz's musical imagination refused to be confined by the practicalities of the opera-house, where what already exists fully in the music for the ear has to be laboriously (and lamely) duplicated on the stage for the eye. Who wants to see the Queen Mab scherzo or Romeo's love-music from *Roméo et Juliette* visually duplicated—if such a thing were possible—in the theatre? And most of the best things in *Les Troyens* have this same appeal exclusively to the musical imagination, which is only distracted by visual representation.

Since *Les Troyens* is as problematic a work as all the rest of Berlioz's output, would it not have been wiser to stage a production in which realism was minimized, where all the grand-opera elements were played down, the eye satisfied with well-chosen spaces, curtains and lighting rather than dazzled by tawdry luxury, and the aural imagination set free to follow Berlioz's musical fantasy? Certainly the supernatural element would have been far more effectively handled in this way; and where Berlioz writes a cantata—as in the final scene of 'La Prise de Troie'—the necessity of inventing non-existent action would have been avoided.

Così fan tutte[*]

Do the individual members of the audiences which have been enjoying the recent revival of *Così fan tutte* at Covent Garden, and in the present Glyndebourne season, feel any of the difficulties and objections that have made this the most debated of all Mozart's operas? Or is this a work which proves unsatisfactory under the analyst's magnifying glass but carries conviction to those who are content to bring to the opera-house no more than the instinctive understanding of human life that they have unconsciously gathered simply by living?

It is fashionable to say that *Così fan tutte* is about sex, which is an inaccurate and misleading, if convenient, way of saying emotional relationships between people of the opposite sex. (*Salome* and *Lulu* are the only operas about sex, as far as I know.) This is obvious; and what is more important, though often overlooked, is that this is an opera about very young people living in a Mediterranean society in which there is a recognized gap between the ideals which society honours by its conventions and the crude facts of everyday existence. We, who have abolished that gap by relinquishing the ideals and the conventions, may find this a hypocritical attitude; but to da Ponte and to Mozart, and all civilized societies prior to our own, it was normal and natural.

Ferrando and Guglielmo, in protesting the fidelity of the girls with whom they were in love, were following an age-old convention, certainly no more foolish than that of young Victorians who disputed and betted on the absolute superiority of their own university, Cambridge or Oxford; and Fiordiligi and Dorabella play their part in the same convention with all the more enthusiasm because they have more at stake. For the young men are certainly not innocent; they have had their experiences of less idealized love-making with girls like Despina, who do not accept the ideals preached by society and only pay lip-service to them in public. But for Fiordiligi and Dorabella, with no sexual experience, love is still a mirage, infinitely tantalizing but only half guessed at, an aura of idealized emotion and only half-recognized desire which may suddenly surround the head of any personable young man who flatters them. There is absolutely nothing unexpected, still less reprehensible, in two such girls, whose very innocence has made them unnaturally susceptible to the idea of love, finding the incarnation of that idea in two different young men within the space of twenty-four hours.

Of course, da Ponte deliberately exaggerates by insisting on the classical unities; and he heightens the comedy by introducing parodies of the *opera seria* in which, as in French classical tragedy, society expected to find its ideals presented in the highest, most uncompromising form and adorned with all the graces that literature and music could provide. In behaving 'like operatic heroines', Fiordiligi and Dorabella were instinctively modelling themselves on the personifications of Constancy or Fidelity whom they had been educated to admire.

The essence of the comedy lies in the disproportion between the grandeur of their gestures and the infinite lability, or 'transferable-ness' of their emotions. They are like little girls who have read about Joan of Arc and imagine that they hear her 'voices' in the whispers that come from the next dormitory, mistaken but touching; and to indulge in moral indignation at the expense of Fiordiligi and Dorabella is as uncomprehending as to punish the schoolgirls for telling lies. Theirs was no more than a 'mock-up' of real love, but one can do well or badly in a mock-up, and they in fact do very well. Fidelity, or constancy, is a quality only demanded in those who sit the real examination; what the mock-up shows is, as it were, the natural gifts of the examinee, interest in the subject, general grasp of the field, and ability to express herself.

It is sexual experience that enables Ferrando and Guglielmo to go through the comedy of disguise (it is unthinkable that the roles should be reversed, and the girls agree to test the men) and sexual experience that accounts for the violence of their feelings when each succeeds with the other's girl. For they have a clear idea of what real 'success' means, whereas to the girls it is still a mystery. Don Alfonso's real cynicism lies in his belief that all women remain schoolgirls like the sisters, that no woman ever grows up, or achieves that emotional stability which is the chief mark of maturity. Like all cynics, he is exaggerating a part of the truth and turning a blind eye to the rest; and no doubt he was in his day a strong opponent of women's education which, by giving women other things to think of beside men, has done more than any other single factor to make his generalization untrue. Despina shares his cynicism, which she finds flattering— an attitude still common among pretty, stupid women—and believing that sexual experience is all that matters in life, she patronizes her young mistresses for lacking it.

Why did this story release a stream of music extraordinary even by Mozartian standards? No doubt da Ponte's symmetrical framework, whose artificiality is part of its charm (though it is the frame, not the picture, that is artificial), answered to the instinct for purely formal beauty, which was as strong in Mozart's musical nature as his instinct for dramatic truth. J. W. Lambert is the first, as far as I know, to mention the name of Jane Austen in connection with *Così fan tutte*, and the comparison is well worth pursuing. But do we raise our eyebrows because *Pride and Prejudice* has not the tragic depth and the painful sincerity of *Anna Karenina*? Or is Mozart to be blamed for the richness of a genius which enabled him to create a comedy of youthful manners as perfect, in an entirely different way, as his (far more puzzling) *dramma giocoso, Don Giovanni,* and pantomime-parable *The Magic Flute?*

11

POLITICS AND THE AVANT-GARDE

Composers and the influence of politics[*]

The meeting of extremes, which is a marked feature of art as well as society today, has puzzled to distraction the ordinary music-lover who hankers for the lost 'middle of the road' which he associates (not always quite rightly) with the great art of the past. Was it not Schoenberg who said that all roads lead to Rome except the *via media*? And that has been the principle of his disciples, though they now find themselves rudely jostled by newcomers on the Left into just that middle way their master declared to be a blind alley. Certainly to an outsider it was an amusing, if puzzling spectacle to see, for example, René Leibowitz, Schoenberg's strongest propagandist in France, concerning himself enthusiastically with the performance of Offenbach's operettas, when these returned to favour in the 1950s; and to witness the gradual building up of Kurt Weill and Charles Ives, two of the avant-garde OK composers, into 'major' figures of music between the two wars.

It is very difficult to distinguish, in what seems to be developments of taste of this kind, aesthetic from political elements. These have often been quite as strong among avant-garde musicians as among Russian party-liners, though similar social and political sympathies have led to diametrically opposed aesthetic conclusions. Anti-serialism, for instance, which counted as a Nazi hallmark in Germany after the war, has until recently been the keystone of Soviet, anti-Western aestheticians. Recent controversy over Pfitzner suggests that his right-wing political philosophy still looms so large that it effectively obscures any final, objective rating of his music; and on the other hand, the social-political attitudes expressed in the works

[*] The *Daily Telegraph* 10 May 1969.

of Offenbach, Weill, or Ives—anti-establishment, anti-war, anti-capitalism, anti-religion, anti-bourgeois, or whatever it may be—certainly account for the atmosphere of benevolence in which their music is judged today.

If Offenbach remains as much performed in Russia as he was until very recently, it is an anomaly which must become increasingly obvious as the Soviet regime grows visibly more like the French Second Empire, Offenbach's original target, and the Second Reich, the achievement of Offenbach's great admirer, Bismarck. It is perhaps worth remembering that in the late eighteenth century Catherine the Great showed her strong, if very partial, sympathy with the French *encyclopédistes* (the European 'progressives' of the day) by introducing into Russia the French *opéra comique*, the musical form most closely associated with social criticism. It has always been 'little' music—parody, *opéra comique*, cabaret song, operetta, revue—that has lent itself to the expression of party political sentiment and social rancour. When political motives have consciously inspired a composer in the big forms—symphony, opera, cantata—the result has always been to diminish or trivialize the music.

How many of the 'revolutionary' works of the 1790s in France have survived? Or the Komsomol symphonies, Five-Year-Plan opera, and odes to Stalin from the Russian 1930s? Even a composer of Shostakovich's stature slackens the rein, lowers the tone, coarsens the thought-texture in his politically based symphonies; and it is hardly possible to believe that Prokofiev's opera *The Story of a Real Man*—where the glorification of war and the patriotic sentiment would today count as out-and-out 'fascist'—is by the same composer who wrote *Romeo and Juliet* or *The Love for Three Oranges*.

Is Beethoven an exception to this rule? I believe not. Neither the planned dedication of the Third Symphony to Napoleon, nor the finale of the Ninth Symphony committed Beethoven to any closer political or social programmes than the idea of 'liberty', which meant for him the destruction of feudalism rather than the establishment of an egalitarian democracy. Neither of these works argues or even implies, his approval of Napoleonic policies in Central Europe or a sympathy with early Socialism, any more than we can argue from Palestrina's Masses a sympathy with the aims and methods of the Counter-Reformation; or from Bach's Passions an approval of Lutheran church policy in Saxony. *Fidelio*, with its French libretto, French *opéra comique* first act, and naïve oratorio finale, is the closest that Beethoven ever came to subordinating his 'big' music to social or political preaching (*The Battle of Vittoria* and *Der glorreiche Augenblick* were frankly occasional, frankly pot-boilers); and is it not just the disparity of these features that prevents *Fidelio* from ranking among Beethoven's very greatest works?

The sign that this is so is that we find ourselves making allowances and adjusting our sights historically with *Fidelio* in a way that never occurs to us during the *Missa Solemnis* or the greatest piano sonatas and string quartets. Beethoven's fundamentally unpolitical attitude can be seen from his accepting a commission to write *Der glorreiche Augenblick* to celebrate the gathering of the Congress of Vienna, the greatest triumph of legitimist reaction, which imposed on Europe for another half century exactly those regimes against which *Fidelio* had been a protest. Beethoven did not foresee this: he was an idealist, not a politically minded man, and he needed the money.

To look for such hard distinctions today, when the trans-valuation of values and the confusion of genres prophesied by Nietzsche have become actualities, is unrealistic. That Nietzsche himself considered both as indubitable marks of decadence is beside the point. What we still need is a 'big' music that is unequivocally human and a 'little' music in which wit and style are not sacrificed to easy popularity.

For People—or 'the People'?[*]

The actual circumstances of the interview with Benjamin Britten published in *Pravda* on 18 March made it appear in quite different lights in the two countries concerned. In Britain it seemed to be a gesture of solidarity with the musicians of a country which had just welcomed a group of British musicians and had shown interest and enthusiasm for much of the music performed, and always for the performers. Mr Britten, it may

[*] The *Daily Telegraph* 30 March 1963.

well have seemed, had every right to express his distrust of 'ivory tower' composers, to emphasize the fact that he himself did not compose for a clique, and to state his belief in 'the artist's social duty—to form, educate and develop [the] people's artistic taste'. But it may be doubted whether he realized the important implications, for Russian readers, of the difference between 'people' and 'the people'; and whether this implication, quite clear in the Russian text, was made clear to him before it was published.

'Art for the people'—not 'for people'— is a slogan that appears in Russian concert halls in any case; but during the British musicians' visit it took on a heightened significance as a result of Mr Khruschev's violent attack not only on contemporary Western art (that was to be expected, perhaps) but on any art that cannot be interpreted as active support for the Communist cause, any 'peaceful co-existence in the ideological sphere'. It may be doubted whether any Englishman who is not a daily reader of *Pravda*—as I was during the time of the British musicians' visit—can imagine the hysterical violence and absolute unbendingness of these speeches of Mr Khruschev's, or the importance attached to them.

When the interview with Britten was published, Shostakovich had already appeared in *Pravda* with a grateful and respectful message to Mr Khruschev on behalf of Soviet composers. It was inevitable, therefore, that when Mr Britten at the end of his interview declared that 'between the arts of our two peoples there are no barriers', readers of *Pravda* would understand that he was in fact subscribing to the full Communist doctrine of art as an instrument of ideological propaganda. English readers might feel sure of his real intent—to express his solidarity with Russian musicians, as human beings and artists, and his belief in art as a means of human communication. What he presumably did not know (though I find such ignorance culpable in an intelligent man in his responsible position) was that Russians allow of no distinction either between the artist's private personality and his public, official capacity as an accredited Communist propagandist; or between humanity as a whole (regardless of race, class, and so on) which interests the Western artist, and 'the people' who are the official concern of the Soviet artist and tie him all too often to the search for an aesthetic

Lowest Common Multiple rather than Highest Common Factor, as Soviet music shows.

The Russians should understand—and we should make it perfectly clear—that whatever may be aesthetic divergences among British musicians, whether they are traditionalists, dodecaphonists, or practitioners of electronic or 'concrete' music, they are united in rejecting the view of the artist's role as being one of a propagandist for any political or sociological creed whatever. In the free world the artist speaks to people, to human beings as such, and he is free to say what he likes as he likes. That much of today's art is experimental only reflects the spiritual crisis through which the whole world (including the Communist countries, of course) is passing. That much of this art is trivial does not distinguish it from the art of other periods which we remember now by a few great names but which were, in fact, often dominated by now forgotten nonentities. The tragedy of the Soviet artist lies in the fact that only in the most exceptional cases can he operate freely, with his own individual face unmasked; and, far worse, that in the vast majority of cases the mask has grown to the face—that after two generations of twenty-four hours' conditioning a day from childhood onward, he is no longer aware of having grown up in the strait-jacket of Communist ideology. The first duty of artists in the West is to force an awareness and resentment of this strait-jacket on Russian artists, since no prisoner will escape if he is unaware of being chained.

How quickly and deeply the varying waves of ideological indoctrination can affect the artist's attitude to his work struck me at the Bolshoi and Kirov Opera Houses. For almost ten years now, the 'cult of personality' has, as we know, been employed as the official explanation of all that was wrong during the Stalinist era; and since Soviet propagandists make no distinction between politics and aesthetics—which are merely regarded as two facets of the same struggle—'personality' has presumably been at a discount in educational establishments, including artistic academies. At least this seems to be a not altogether fantastic explanation of the fact that, although I heard several excellent singers in a number of good performances, not one struck me as projecting a role with the whole force of his or her personality. The performances were safe and correct and left the listener—or at least me—tantalized by the feeling of unused potentialities. The most noticeable instance was that of Shtokolov, who is the best Boris at the Kirov Opera House in Leningrad and certainly has a magnificent voice and a fine stage presence. To say that his performance gave the impression of conscious inhibition would be an exaggeration. It was rather as though his personality were muted by some instinctive avoidance of complete self-expression, an instinctive cautiousness that made him first and foremost a good colleague unwilling to 'hog' the stage (as any Boris should) or to assert freely the claims of the character that he was impersonating.

Even among the top-ranking Russian artists who have been received so rapturously in the West, it would be difficult to mention more than one, or at the most two, whose performances are not merely technically magnificent but great individual interpretations. Rostropovich certainly and, within the limits of a somehow maimed personality, Richter; but neither Oistrakh nor any of the singers. Those who saw Vishnevskaya's Aïda at Covent Garden may have had the same impression as I—of a 'muted' personality escaping full dramatic expression by a mixture of ham acting and careful vocalization.

When Theatre takes over from Music[*]

'Devils are not to be believed even when they speak the truth.' This ambiguous motto, which Penderecki chose for his opera, *The Devils*, has often recurred to my mind since it was explained to me that any hostility that I might express towards the opera would inevitably be discounted as built-in prejudice; and that indeed I could hardly find valid grounds of objection to the work of a composer described as a 'fervent Roman Catholic'.

In his *Musical Criticism*, which appeared just fifty years ago, M. D. Calvocoressi amused himself and his readers by quoting a form of personal dossier that J. M. Robertson wished every critic to complete, in order to inform readers of his natural, or 'built-in', prejudices. Robertson called it 'a confession of faith, bias, temperament, and training'. Although it is too long to quote here and its terms of reference have in many cases been either superseded or confused, it includes a number of what at

[*] The *Daily Telegraph* 17 November 1973.

that time would have been considered outrageously personal questions. Today it would not be considered enough, in such a context, for the critic to disclose his age, as Robertson required; he could reasonably be asked to supply details of his social origin, sexual tastes, nationality, religious orientation, and skin pigmentation.

These disclosures, however, would still not take into account either the high incidence of guilt and self-hatred among intellectuals or the shame that even the old, let alone the middle-aged, often feel at finding themselves unable to agree with the young. The public prints and the media are crowded with public schoolboys desperately ashamed of their educational and social advantages; with Christians pathetically anxious to assure their unbelieving friends that theirs is of course a God-less religion; with Englishmen obsessed by the evils of their imperial record; and with Jews caught in the web of their own masked anti-Semitism.

This form of self-hatred is a phenomenon that has grown enormously since Robertson's, and even since Calvocoressi's day, and it accounts for a great deal of the violence and bitterness of our controversies. In fact, one of the most important questions in any dossier of this kind would now be: 'Do you belong to the class of birds that at adolescence leaves the nest naturally and returns to visit the parent-birds at regular intervals? Or to those whose leaving is unnaturally early/late and stormy, and who feel a compulsion to foul the nest and bespatter the parent-birds before abandoning both for good?'

Only a fool can doubt that the answers to these questions would indeed throw a light on the critic's aesthetic judgements. But have we unconsciously gone so far along the road to determinism that we believe the attitudes of any man, let alone a critic, are simply the sum total of these answers, and can thus be obtained by plotting a kind of parallelogram of forces? If so, we must accept the paradoxical conclusion that the only subjects on which a man's opinions are worth hearing are those quite removed from the sphere in which he has been born, reared, and trained—and therefore, so the argument runs, hopelessly twisted, or brainwashed.

The equating of informed conviction with prejudice, which is very common, implies having despaired of reason, and resigned oneself to the instincts of conflicting herds. I should like to maintain that although I am a white Christian of sixty-three and was educated at a public school, none of these circumstances invalidates my opposition to the showing of torture on the stage, as was done in *The Devils*. That opposition is based on a rational understanding of human nature, as observed quite as much in myself as in others; and on an equally reasonable concern for the quality of human life. Neither of these is the preserve of any age group, social class, religious creed, or colour.

Mr Penderecki calls his opera a protest against intolerance, and we must believe that it is thus that he conceived it. In the same way torture scenes in the cinema are often justified, by the plea that they only show things that are happening all over the world today. But if this plea is in good faith, it is extremely naïve; and who in fact would care to assert that it is moral indignation that fills cinemas for films showing scenes of torture? Protesters, like satirists, should think twice before depicting the vices they profess to castigate. Nobody ever read Juvenal's flayings of Roman society for their moral indigation; their attraction has always lain in the wit and vividness of the descriptions.

Mr Penderecki failed to take into account the fact that the spectacle of human degradation, humiliation, and physical torture only revolts that part of our nature which the efforts of parents, schoolmasters, and eventually ourselves have managed to raise above the level of the jungle where 'unregenerate' man, however civilized, very soon finds himself at home. Can Mr Penderecki still believe that Hitler's concentration camps were staffed by people essentially different from himself? Or that, given the necessary pressures, the public opinion, and a misleadingly gradual introduction to 'Draconian methods', he could have not found himself equally callous to human suffering and eventually relishing it? If so, he need only watch a 'child of nature' tear the wings off a fly or torture a kitten.

It is to this instinct, which is unfortunately human rather than animal, that torture scenes appeal; and there is a tragic, unconscious hypocrisy in those who protest against the use of judicial torture in real life, and yet enjoy scenes of simulated torture in the theatre. In fact, it is Mr Penderecki's common sense

and human understanding of human nature that I am questioning, rather than the fervour of his Roman Catholicism,
something perfectly compatible with an absence of both
common sense and human understanding.

12

COMPOSERS AND MUSICIANS

Clement Harris[*]

Before the days of demonstrations, when young men who felt strongly about a foreign cause by-passed their Governments and went themselves to fight, Clement Harris was killed at Pente Pigadia in Epirus, on 23 April 1897. Harris was a philhellene, who raised an armed force in Corfu in the first days of the Greco-Turkish War, landed on the mainland of Epirus, and was wounded a few days later, and then finished off by a janissary, outside the Turkish fort of Pente Pigadia, three months before his twenty-sixth birthday.

In his diary he speaks of his enlistment as 'the least a man of honour can perfórm towards a country which, crying for liberty in the name of the Cross, has been insulted and thwarted by each so-called civilised Power successively.' Naïve self-dramatization in a pompous late-Victorian style, perhaps; but all young people dramatize themselves, very properly, and most of those who are going to do much in the world go through a period of priggishness, veiled or open. Harris certainly does not need to apologize to the Vietnam demonstrators of today.

His death earned him, at the Empress Frederick's special request, a memorial in the Anglican church in Athens; but how did he come to rate an entry in the latest edition of *Grove*, or indeed a place in this column? A *feuilleton* devoted to Stefan George's poetry, published in the *Frankfurter Allgemeine Zeitung* on 5 December, put me on to Harris's track. He, it seems, was the 'Clemens' who figures with Nietzsche, Böcklin, Leo XIII, and Elizabeth of Austria (awkward table-companions)

[*] The *Daily Telegraph* 4 January 1969.

in the 'Zeitgedichte' of George's *Der siebente Ring* (1907); and for a short time he seemed to be one of England's most promising composers.

He was the son of a rich export merchant and was born at Wimbledon in 1871. At eighteen he left Harrow to study music at the Frankfurt Conservatory, in the first place under Clara Schumann, who seems to have made a remarkable pianist of him. Schott published four concert studies of his—*Ballade, Il Penseroso, L'Allegro*, and *Lied,* the transcription of a song by Peter Cornelius. From 1893–6 Harris studied instrumentation (and perhaps composition?) under Philipp Wolfrum in Heidelberg and composed a symphonic poem, *Paradise Lost*, which was accepted for performance in Germany but withdrawn by the composer at the last moment and only performed after his death (Birmingham, 1905). His *Festmarsch für grosses Orchester*, on the other hand, was performed not only in Germany but—according to Werner Helwig's article in the *Frankfurter Allgemeine Zeitung*—at 'a court betrothal ceremony at Westminster', when the composer's name first came before the British public and raised high hopes for his future.

In Frankfurt Harris made friends with the art historian Henry Thode, whose wife Daniela was a daughter of Hans von Bülow and a stepdaughter of Wagner's. Through her he met and became a close friend of Siegfried Wagner, then an architecture student at Karlsruhe. In February 1892, the two young men embarked in one of Harris's father's boats for a cruise which took them as far as Japan and the Philippines; and it was during this voyage, on Easter Day, that Siegfried finally decided to abandon architecture and join his mother in the administration of Bayreuth and to become himself a composer. That choice was not an altogether happy one, though perhaps inevitable; and of Siegfried's twelve operas hardly even a memory remains. But according to Werner Helwig, Harris's death prompted Siegfried to write a symphonic poem, *Glück* ('Good Fortune', a strange and perhaps significant name for a funeral ode), which is not mentioned in the official lists of his works and was only given a private performance in 1923 but 'may perhaps rate as the only work which sprung from a region outside his father's overwhelming influence and is worthy of revival'.

How Clement Harris came into the Stefan George circle is not clear*. Werner Helwig's casual suggestion that his philhellenism 'marks his erotic proclivities and his claim to George's friendship' may be based on firm evidence. If not, it seems a very flimsy argument. The poem which he inspired is a characteristic cross between a sculptural Parnassianism and apocalyptic hymnography. The opening is fine:

> Als ihn im kampf des Türken kugel warf
> Am ölwald von Epirus: blieb der kummer
> Nur uns um dieses blumenschweren frühlings
> Zu rasche welke . . .

(When in battle the Turkish bullet struck him by the Epirus olive grove: the grief was ours alone for the too-early fading of this flower-laden spring. . . .)

Perhaps the BBC Music Programme will give us the opportunity of hearing Harris's piano pieces and *Festmarsch für grosses Orchester*, and Siegfried Wagner's unique *Glück*.

Edward Dent†

'From an artistic point of view most of old music is much better forgotten, and it is only civilization that has forced us to remember what we ought, in the nature of things, to have forgotten.' These sharp words were not written by a bright young composer indignant at finding his own music less often performed than Beethoven's, but by a witty and erudite professor of music best known for his work on Alessandro Scarlatti, Handel, and Mozart.

Edward Dent was in fact fifty in 1926 when he wrote the long essay which contains these and many other lively—and, as it turned out, sometimes prophetic—opinions. Originally published

---

* I am indebted to Dr. R. C. Ockenden for the following note: 'A study by Claus Victor Bock, entitled *Pente Pigadia und die Tagebücher des Clement Harris* (Castrum Peregrini Presse, Amsterdam, 1962) gives a full account of Harris's brief life and its contexts, based on a close study of his diaries, and a full list of his compositions. It also explains his contact with Stefan George, which led to the writing of George's commemorative stanzas, "Pente Pigadia". In a footnote to his book, published in *Castrum Peregrini* LXXV, 1966, Bock advances the arguments for supposing a meeting between George and Harris in August or September 1896; but there is no suggestion that Harris ever "came into the Stefan George circle".'

D. C.

† The *Daily Telegraph* 15 January 1966.

in the 'Today and Tomorrow' series (which also included W. J. Turner's *Orpheus*), Dent's *The Future of Music* has recently been reissued as a paperback and is well worth reading for its combination of learning about the past and clear-sightedness about the present of music.

Although a scholar by profession, Dent was acutely aware of the danger that exclusive preoccupation with the past could represent to a musician—hence his aggressive tone in speaking of popular indifference to new music and the enthusiasm with which he worked for its promotion. The emotional stimulus of this work of propaganda was in his case a strong nonconformist bias such as is not uncommon in this country among intelligent members of the privileged establishment—Bertrand Russell is the most obvious example. Edward Dent reacted violently, early, and for life, against the late Victorian world in which he was brought up—Yorkshire gentry, Eton, and King's; and, of course, in the middle distance of all his youth, the Church of England. He grew up into a Continental-type anti-clerical who lived at Cambridge in a way almost indistinguishable from a worldly eighteenth-century French or Italian abbé, and a democrat whose aristocratic self-confidence and bitingly sardonic tongue must have frightened to death many of the young men with whom he was anxious to place himself on a level.

These prejudices can easily be traced in his mistrust of ethical fervour ('reverent pomposity') and religious exaltation ('hard to distinguish from rhetorical pretentiousness') in the work of nineteenth-century composers, especially Beethoven. He rightly regarded it as one of the tasks of all new musical movements to destroy the 'tyranny of association' which often causes the listener to concentrate his attention not on what the composer wrote but on the store of emotions that familiar music has laid up in his own sensibility. But is he not a victim of his own associations when he says that the great works of Beethoven's last period lead us into a 'metaphysical labyrinth' and suggests that 'the reality of music lay . . . in sounds—or rather relations between sounds—never actually heard at all, but induced in the perceptive faculty by association'? Here he is surely reacting against a view of Beethoven inculcated, perhaps, by Charles Wood or Stanford (both of whom taught him) rather than expressing his direct experience of the music itself. And it is hard to know exactly what he means when he makes this 'ethical' or 'associative' attitude to music responsible for lowering standards of singing.

In the music of his own day he defends 'intentionally destructive' music (whom had he in mind? Les Six?) but dismisses such 'emetics' as forming by themselves a poor diet. He seems perhaps to have been casting a long glance into the future when he wrote of a cult of the 'mysteriously fragmentary', though he dismisses this as a new form of the sentimentalism which was always to him the chief enemy. It was no doubt traumatic memories of his mother's drawing-room that prompted his shrewd observation that the 'refinement introduced into music by Couperin, Boccherini, and the lesser Mozart' (perhaps really the 'galant' style of the rococo in general) was to have a disastrous effect on the vitality of the art. Vitality was what he prized wherever he found it, and particularly in the new music of his own day. He could be merciless about music of the past, particularly if he could discover in it any reflection of his pet aversions—as in the case of the fifteenth-century Netherlanders whose 'poverty of melodic invention, passionless indifference to sensuous beauty, and rigid obedience to rule' seemed to him 'monkish virtues'.

He even carried his nonconformist obsession into his theorizing on the nature of musical enjoyment. In listening to the music of the past, he says, it is the 'moments of adventure' that move us, when we join the artist in perceiving intuitively and directly something that we know to be true and beautiful 'although it is not consistent with the conventional principles on which the art is based'. Such pleasure is in any case confined to well-trained and highly conscious musicians; but it may be doubted whether even they enjoy chiefly, let alone exclusively, the moments when a composer is introducing a personal innovation into a generally accepted language. This is in fact a purely romantic attitude to music, since it equates interest and vitality with the expression of personality.

It is interesting to find here an echo of the still topical debate on 'aesthetic emotion' which Clive Bell and Roger Fry had started with their *Art* (1914) and *Vision and Design* (1920). To achieve this, Dent writes, we have first to learn to appreciate music rationally; but the moment of aesthetic emotion will only be granted when we are suddenly faced with some irrational moment as the nonconformist 'adventure' which I have just described. Such questions of aesthetics seem remote today, when many of what were still in 1926 unquestioned elements are no longer present in our musical experience. Dent never assumed in this essay the mantle of the prophet. His style, like his character, recalls by its dry clarity the eighteenth-century French *philosophes* and their neat and reasonable universe. Re-reading this essay I was reminded of what George Santayana said about the French mind—that it was an exquisite medium for conveying such things as can be communicated in words. 'It is the unspoken things of which one feels the absence or mistrusts the quality'—and the same is true of Edward Dent's understanding of music.

Hubert Parry[*]

The death of Sir Hubert Parry in the autumn of 1918, some six years after Debussy, is a convenient milestone marking the end of a long period during which English composers looked abroad for their models and produced provincial versions of accepted Continental styles, anything up to fifty years after these styles had been superseded elsewhere. Handelian oratorio, for instance, was being written in this country almost up to the time when Mendelssohn took Handel's place; and Mendelssohn and Gounod between them accounted for a very large proportion of the serious music that was written here almost to the end of the nineteenth century.

The revolution attributed to Parry and Stanford in the 1880s (dating from Parry's *Scenes from Prometheus Unbound*) was really more a reduction in the time-lag separating this country from the rest of Europe—the substitution of the living Brahms as a model for Mendelssohn, already thirty years dead—than the discovery of an English composer with an unmistakable voice of his own. This was only to appear with Elgar; and the gulf that separated the two men's music was made very clear in a memorial concert of Parry's music last week, when his *Symphonic Variations* and the oratorio *Job* were performed. What distinguished these works from hundreds of other

[*] The *Daily Telegraph* 9 November 1968.

Brahmsian offerings was the sensitive handling of English words, but also the clumsy, almost amateurish quality of all the orchestration, and not a little of the actual composition.

Parry was a great influence in English music, but less as a composer than as an administrator and a raiser of standards, not only in teaching but in the status enjoyed by music in the public eye. Although Sullivan was unquestionably the more gifted musician, he was the son of an Irish bandmaster and wrote operettas, while Parry was a country gentleman's son (and married to a peer's daughter) who wrote oratorios, and there could be no question to the late Victorian mind of their relative importance. The combination of moral elevation, quirky humour (*The Pied Piper of Hamelin*), high-minded agnosticism, and good breeding made Parry an irresistible paragon to the Victorian establishment and stirred in Elgar, the infinitely more gifted musician, an agonized resentment and perhaps an instinctive desire to imitate—to conceal the sensitive music-shop boy behind the country gentleman's mask.

Parry's literary taste and much of his well-bred amateur's attitude to music, his preference for choral music and 'manly' themes and his Britisher's distaste for the opera were all inherited by Ralph Vaughan Williams, who incarnated the nationalist movement in music which flourished here a good half-century later than elsewhere and finally disappeared only with the Second World War.

It is very easy to be superior about the past, and particularly that closer past that is still old-fashioned rather than antique. Perhaps it would help matters if we spoke, as in the case of the East, of the Near, the Middle and the Far Past, and with a comparable absence of implied value-judgements.

It is a salutary exercise to remember that in the huge majority of cases composition is always based largely on the imitation of some model; and that more than ninety per cent of all compositions lose all interest when the music on which they are modelled itself goes out of fashion. We have been speaking of music modelled on Handel, Mendelssohn, and Brahms by composers in whom the faculty of imitation was stronger than the faculty of creation. But we should bring the story up to date and remember the music modelled by similar composers on Bartók and Stravinsky between the wars, and on Schoenberg,

Webern, and Messiaen since 1950. There can be no reasonable doubt that much of this, including many works that are still admired, will soon seem quite as dead as Parry's *Job* and *Symphonic Variations* seem today.

It is not even necessary to be very old to be aware of this process happening before one's eyes. Any music lover of fifty in this country will remember the enormous enthusiasm which greeted every single work by Delius, Vaughan Williams, or Sibelius between the two wars; and how, when these local gods fell, every work by Stravinsky or by a member of the Second Viennese School (and even, in some cases, by their pupils) was hailed as an unassailable masterpiece. Time has already begun the process of weeding, the undermining of that orthodoxy and the quiet smuggling back into the repertory of judiciously selected examples of the orthodoxy before last. Barriers once thought insurmountable are falling in the department of musical taste as elsewhere, and today we see no objection to a 'permissive' programme in which, say, Delius, Webern, Tchaikovsky, and Boulez share the bill. And the moral is that -isms are out and the quality of a composer's musical personality is all that counts. It always was, of course, but the fact was obscured by an exclusive partisan spirit that now seems the only thing that is really 'old-fashioned'.

Modest Mussorgsky[*]

It is a remarkable fact that the barrier of language has not prevented the great Russian novelists of the nineteenth century from conquering Western Europe but has proved more formidable an obstacle to the full appreciation of the most original of Russian composers. Mussorgsky, the hundredth anniversary of whose death fell on 2 March, belongs among the great Russian artists who will always be rated more highly by their own countrymen than by the rest of the world. He shares this position with Pushkin and Gogol among writers, and with Glinka among musicians, and his case is more similar to Gogol's than to that of his fellow-composer. Pushkin was the supreme master of the Russian word, creating in his works a new language and with it a new sensibility that pales in any

[*] The *Daily Telegraph* 21 March 1981.

transcription, and all but vanishes in any translation. Glinka's gifts as a composer were incomparably inferior to Pushkin's as a writer; but he too invented a musical language, and with it a sensibility, which reflected the tone of the society for which he wrote—brought up by French governesses and German tutors, spending often long periods in Western Europe where they were as homesick for Russia as they were bored when they returned there. This splintered personality is in Glinka's music, as it is in much of Tchaikovsky's and Stravinsky's: it is the world of Turgenev's novels.

Gogol and Mussorgsky belong to a world that is different because it is hardly touched by Western modes of feeling or Western emotional conventions. Both these great artists delighted in the exploration and depicting of what was specifically Russian in the world of their experience; and the vehicle through which this communicated itself most clearly and most fully was the Russian language. To translate Gogol is comparable to transcribing an orchestral work for the piano: in each case the original text must be rethought rather than simply translated into the new medium—and yet its specific flavour will still be lost.

If all the important music written by Mussorgsky is vocal, it is because the Russian language itself was, in one sense or another, always his inspiration. Even the handful of orchestral and piano pieces by which he is remembered have a visual scaffolding, for to think consecutively and coherently in musical terms was impossible to him. This fact allies him to his successors rather than to his classical predecessors. Mussorgsky's instability, his inability to order his own existence, even his drunkenness, were interpreted by his passionate admirer Shostakovich as an integral, if unconscious part of his protest against the ignominies and indignities, the cruelties, the squalor, and the hopelessness of life as it was for the great majority of the Russian people. He was at the opposite extreme to the intellectual 'dissidents' of the day—Belinsky or Herzen—who did not share the conditions against which they protested. Mussorgsky shared suffering, suffered vicariously, and expressed his protest in the parable-like language of his music and his instinctive refusal to conform with convention, to 'live properly'. His protest was that of the 'holy fool' of Russian tradition, the Simpleton who at the end of

Act IV of *Boris Godunov* openly taxes Boris with the murder of the Tsarevich and yet is spared. *Boris Godunov* had an absolute fascination for Stalin, and Shostakovich recounts the thrill of horror with which he himself used to hear the stage Boris excuse himself with the words: 'It is not I . . . it's the people . . . the people's will'—as though Stalin himself were speaking.

There are now three versions of *Boris Godunov*, all slightly different in musical character and dramatic emphasis. Rimsky-Korsakov's smartened and conventionalized recension of the composer's original (itself of doubtful authenticity in some respects) has been replaced in the Soviet Union by the re-orchestration in which Shostakovich attempted to achieve what he believed to be the composer's own ideal—'a sensitive and flexible orchestra in which the vocal line is surrounded by the subsidiary voices of Russian folk song'. The old dispute as to whether Tsar Boris or the Russian people is the hero of the opera reveals a totally Western misunderstanding: the subject of *Boris Godunov* is the relationship between absolute authority and those over whom it has power. Once this is realized, the succession of vignettes resembling a cinematic film rather than a stage play appears as the only possible form in which the drama could be presented. What holds it together is Mussorgsky's music and the determining factor in that music is the Russian language, with its characteristic rhythms, cadences, and rich alternation of liquid vowels and rough consonantal groupings, its Byzantine assonances on the one hand and folk-song affiliations on the other. The combination of Gogol and Mussorgsky in the recitative-opera *The Marriage* and in *Sorochintsy Fair* (both unfinished) is so language-dominated as to be impracticable for foreigners to perform; and although *Khovanshchina* has achieved a certain popularity in Western Europe during the last fifteen years, little of the music has the compelling power of *Boris Godunov* and the provincially Russian interest of the drama is impossible to make general.

Apart from *Boris Godunov*, it is Mussorgsky's songs that keep his reputation alive. Many of these are again vignettes—dramatic cut-outs of individual Russian types against vividly dramatic backgrounds—and again the character of the song depends in almost equal proportions on text and music. The observation and the purely human appeal of the 'Nursery' songs are not without hints of the sentimentality all but inevitable in any nineteenth-century artist's handling of children, and this probably contributes to their popularity. More important musically are the two song cycles written in 1874–5, both with texts by A. A. Golenishchev-Kutuzov, a kinsman of Napoleon's opponent in 1812. In *Songs and Dances of Death*, Mussorgsky uses a kind of arioso recitative to evoke four different faces of death as it presents itself to a child, a young girl, a drunken peasant, and lastly those killed in battle and answering the final roll-call of Field-Marshal Death—a pendant to Glinka's earlier setting of Zhukovsky's *The Midnight Review*. In the other cycle, *Sunrise*, the interest of the poems is more broadly lyrical and less specifically Russian; and the last of the six songs, 'On the river', is perhaps both the most finely conceived and musically the most self-sufficient of all Mussorgsky's non-operatic pieces. Yet here too the music is inextricably interwoven with the Russian text, and to translate it is to mutilate the song. Thus the language barrier still presents an obstacle to the full recognition of Mussorgsky's stature and is at the same time an indication of the limiting effect of language upon music. It is significant that Mussorgsky's name is not mentioned in Alfred Einstein's *Greatness in Music*, written in 1941 but from the classical German nineteenth-century standpoint—from which Mussorgsky could only appear as a Slavonic sideshow in the great musical spectacle.

Alexander Dargomyzhsky[*]

There is little doubt that Alexander Dargomyzhsky's name has been a real historical handicap to him. Unable to translate it like the eighteenth-century Bohemian Myslivecek, who called himself Venatorini, he would have been well advised to forestall Hollywood and call himself Dargo, so that both admirers and enemies could at least refer to him without embarrassement. As it is, an unpronounceable name, coupled with the fact that the words, to which he decided to give the primacy in opera, are Russian words and therefore meaningless to most Western Europeans, have told seriously against the spread of Dargomyzhsky's music. Outside Russia he remains little more than a history-book

[*] The *Daily Telegraph* 8 February 1969.

figure; and when Beecham put on his *Rusalka* in London during the nineteen-thirties, even Chaliapin as the Miller failed to interest the public, though today's more inquisitive opera-lovers would certainly show more discrimination.

Dargomyzhsky, the centenary of whose death fell on 17 January, had a characteristically Russian background. His father was an aristocratic by-blow of the Ribeaupierre family, his mother a Princess Kozlovskaya; and the whole background of his life was urban St Petersburg rather than some remote country estate. He was an enthusiastic composer in his teens, already attempting string quartets as well as piano pieces and songs; but it was in 1834, when he was twenty-one, that his first meeting with Glinka led to his studying theory and giving his random attempts at composition a more conscious form and direction. It was, of course, a case of the blind leading the blind, from the point of view of any professional musician in Western Europe, for whom the notes that Glinka had made from his studies with Dehn would hardly pass as a substitute for formal musical education. In Russia, however, Dargomyzhsky could soon pass as a professional.

He was already dominated by his instinct for words, and his early songs show a natural sense of the colours, contours, and rhythms of Russian speech. Even his early, Frenchified opera *Esmeralda* contains clearly Russian popular traits in genre romances and town dance. Between finishing *Esmeralda* in 1839 and its first performance in 1847, Dargomyzhsky visited Paris, where he was more interested in the racy vaudevilles than in conventional operas, while in the law courts he could observe the uninhibited quirks of human nature. He found the 'craftsmanship and intellect of *Les Huguenots* incredible, but no craftsmanship or intellect can imitate the human heart.' In the courts he could 'witness the very actors in a live episode and follow the development of the action'; and, like any subject of Nicholas I, he must have heard with incredulous delight 'the caustic comments about littérateurs, new plays, Ministers, etc.' which formed the chief attraction of the vaudevilles.

This close coupling of music with everyday life, which was something new in nineteenth-century music though very familiar in Renaissance Italy, found expression in the dramatic monologues that Dargomyzhsky wrote on his return from Paris. For his texts he turned from Pushkin to Lermontov and Koltsov, and his comic monologues include one with peasant words which he himself took down ('The Fever') and arrangements of gypsy songs generally considered beneath the notice of serious composers. In his next opera, *Rusalka* (1856), Dargomyzhsky not only used individual folk-songs but, like Stravinsky, himself composed in the folk-song manner, so that the whole work has a stylistic unity new to Russian music. The national element is particularly noticeable in the big choruses; but Tchaikovsky pointed out that the 'inimitable originality' of *Rusalka* lay in the dramatic recitatives.

Dargomyzhsky was a teacher of singing (though he only accepted women pupils) and he was therefore in a good position to understand exactly the problems of presenting an operatic role and the importance of verbal enunciation. His original solutions can be found in his last work (left unorchestrated at his death), a word-for-word setting of Pushkin's Don Juan drama *The Stone Guest*. This includes a telling scene in which the Don surprises Donna Laura, an old flame, with a successor whom he kills; and a macabre scene in which Don Juan disguises himself as a monk and declares his passion to Donna Anna at the grave of the Commendatore, who is not Anna's father but her husband.

The whole atmosphere of Pushkin's piece, which Dargomyzhsky catches with uncannily economical accuracy, is dry yet nightmarish, fleeting yet precise, revealing by single strokes a world in which violence and debauchery are taken for granted and lie very near the surface of everyday life. St Petersburg of the 1820s was certainly nearer to sixteenth-century Madrid than the comparatively civilized and orderly Vienna of Joseph II.

To Don Juan's ugly logic in refuting Donna Anna's instinctive reaction that he is mad, Dargomyzhsky imparts a melodic angularity and a harmonic ambiguity that themselves suggest something twisted, if not actually insane. Again, the whole-tone motive, thundered out in the brass when the Commendatore appears, already forms a creeping bass beneath Leporello's realistically breathless words of invitation to the statue.

The Stone Guest is unquestionably Dargomyzhsky's most original work, a uniquely successful essay in a style which cannot rival, still less replace, music-dominated opera but forms a legitimate parallel cul-de-sac. Like Mussorgsky's *The Marriage*, for which it provided the pattern, it all but completely defies translation, as must any work so consciously and scrupulously founded on the rhythms and intonations of a language, and enriched by so little musical development. Dargomyzhsky's greatest historical achievement is to have prepared the way for Mussorgsky, in whose works music is absolutely paramount whatever his theories may have been. In Janáček's operas, too, we miss the speech-realism, but the music alone is more than enough to hold our attention. In *The Stone Guest* the normal operatic experience is reversed and the nakedness of the musical phrase is only clothed by the body of the text. Yet Dargomyzhsky was imposing a self-denying ordinance on himself when he forswore symmetrical melodic forms in his opera, for his songs show him to have been, like Gluck, a quite exceptional melodist.

François Couperin[*]

After the Bachs, the most distinguished family dynasty in musical history is probably that of the Couperins. Between 1655 and 1826 the church of St Gervais in Paris had eight organists, only one of whom was not a member of the Couperin family. As firmly rooted in French musical tradition and institutions as the Bachs in those of Germany, the Couperins' music enjoyed more respect than actual performance outside France, at least during the long period of German's musical hegemony, which began very soon after the death in 1733 of the family's greatest representative, François Couperin 'le Grand', and lasted until recently. The French scale of composition, French concern with detail, and (in the case of his keyboard works) the importance and complexity of the ornamentation and the whimsical literary taste of the titles all combined to give this music an air, at first of old-fashioned courtliness and then (by false analogy) of triviality.

François Couperin, who was born three hundred years ago on 10 November, was indeed a court composer, in the sense that he became one of the organists of Louis XIV's Chapel Royal at the age of twenty-six, and wrote much of his harpsichord and chamber music for what may roughly be described as court

[*] The *Daily Telegraph* 26 October 1968.

performances. But in the France of Louis XIV outstanding artistic ability was certain, sooner or later, to attract royal attention and patronage; and even the artists who enjoyed neither still thought in no other terms and aimed at no other listeners than those of the court. This did not, of course, prejudice his contemporary admirers, who included Telemann, against his music; and now that the standards and ideals of the nineteenth century have receded into historical perspective, it no longer worries today's listener.

It is not then Couperin's style that prevents his music from being more widely known, but the forms in which that music is cast. The early chamber music, mostly in trio sonata form, and the church works, vocal and instrumental, clearly reflect the Italian baroque taste. In the motets and *Tenebrae Lessons*, Wilfrid Mellers, Couperin's greatest English champion, finds French qualities that are equally present in Fénelon's prose and Fauré's *Requiem*; but none of this music can, of its very nature, reach or make much appeal to the general public. For the great majority of people, musicians or otherwise, Couperin's name will always be associated with the harpsichord, which commands a growing public, and with the two hundred and twenty odd pieces that he wrote in the twenty-seven *ordres*, divided into four 'books'.

It is impossible to generalize about the character of this large output. The conventional description of Couperin's harpsichord pieces as highly-wrought miniatures comes hopelessly to grief with the eighth *ordre*, which includes allemandes and courantes worthy of J. S. Bach for grandeur and complexity, not of surface but of harmonic design, and culminates in a superb and tragic passacaglia. Again, the bagpipes, fiddlers, and street-songs that are echoed in the 'Fastes de la Grande et Ancienne Mxnxstrxdxsx' [*sic*] of the eleventh *ordre* make it impossible to speak of Couperin's art as exclusively aristocratic. Dance music, either simple or sophisticated, and echoes of lute pieces from the preceding century account for the shape, rhythm, and layout of many of the pieces. But there are echoes of Lully's operas, overtures as well as passacaglia, in the twenty-third and twenty-fifth *ordres*; and in many of the later pieces the bold harmonic experiments, introduced as easily and as apparently casually as Chopin was to do a century later, stand in the greatest possible contrast to the very simple harmonic idiom of the earlier dance movements.

All Couperin's harpsichord music is marked by a profusion of delicate ornamentation which he himself insisted was organic. 'My pieces must be played as they are written and . . . if this is not done accurately, they will never impress persons of true taste.' Even in the eighteenth century, Dr Burney complained that his pieces were 'crowded and deformed by beats, trills and shakes'; but no one who has heard this music played by an artist who has really mastered the highly ornate style, and to whom it comes as second nature, will echo his complaint. There is an inexhaustible fund of nuance, comment, and quiet drama in this pure finger-play; but it needs a consummate performer as well as a quite unusually attentive listener, and Couperin is still waiting for the performer who, like Ralph Kirkpatrick in the case of Domenico Scarlatti, will capture the imagination of the public by his performances.

Couperin's titles provide material for a study of their own, and one that would provide a key to the intimate, local, personal element which plainly played a large part in Couperin's art. Here bird-pieces alternate with sketches of local or personal character; topicalities with whimsical jokes; personifications of virtues and vices with straightforward dances; and the whole is interspersed with deliberate mystifications and puns. But we need attach little more importance to Couperin's verbal fantasies than to the titles which Debussy attached, as an afterthought, to his piano preludes. In both cases the music is self-sufficient, and Couperin's titles have misled serious, well-meaning foreign musicians who have no knowledge of the period and little understanding of French humour. For these pieces of Couperin's are French in the sense that only domestic, provincial art untouched by Parisian cosmopolitanism, was to remain French much longer. It is easy to forget that even the never-ceasing round of pomps at Versailles was only a thin facade masking a France still overwhelmingly rural, with local patriotisms and usages often much stronger than the official nationalism and a humour in which the *esprit gaulois* was a far stronger ingredient than the verbal wit that looms so large in the official literature and memoirs of the day.

Couperin's art combined the fine workmanship and much of

the sensibility demanded by the court with this older, deeper, far more robust, and more diversified culture. His harpsichord pieces provide a unique series of vignettes, each of them perfect in balance and construction though they differ in importance and elaboration as they do in character. After three hundred years hardly as large a proportion of them is familiar, even to musicians, as of Schubert's songs or Haydn's symphonies and string quartets. Here is an opportunity for exploration with a certainty of no dud discoveries and a very high proportion of treasures, the finest tercentenary celebration that can be imagined.

Maurice Ravel[*]

Maurice Ravel, the centenary of whose birth fell on 7 March, probably represents in many people's minds the quintessential French composer, and certainly his music has many of the qualities which distinguish much French art. The scale is modest, the workmanship impeccable, the taste refined without affectation; and the whole is determined by clarity of intention and texture, precision of ear, and an unfailing sense of proportion. It is an art that in many ways recreates the sober values of Chardin and the fantasy of Watteau in a twentieth-century setting.

Metropolitan France was not, however, the country of either of Ravel's parents. His father was a Savoyard and his mother a Basque, and genealogically Switzerland and Spain might have as good a claim to him as France. The belief that the origin of his father's family was Jewish is based on a mistaken etymological derivation of the name Ravel from 'Rabbele', the little rabbi; and this belief was naturally strengthened when his *Deux Mélodies hébraïques* seemed to suggest the deep personal involvement which the composer almost invariably concealed behind a formal mask.

All Ravel's music springs from the instinct of impersonation, the desire to evoke and to incorporate himself in figures and landscapes totally remote from everyday life and from his own self. If this attraction to the exotic may be considered a 'romantic' trait, his horror of emotional display or of any hint of

[*] The *Daily Telegraph* 15 March 1975.

autobiography in music gave him an immediate sympathy with
the anti-romantic reaction. He detested the music of Berlioz,
which is indeed the exact antithesis of his own in every way. The
aliases behind which Ravel composed reveal as much of his
character as the music itself. His deep devotion to his mother—
and after her death to the brother with whom he made the rest
of his life—is shown in the two worlds inhabited by the greater
part of his music: Spain and the nursery. Spanish inspiration
accounts for whole works ranging from the early *Pavane pour
une infante défunte* through the *Rapsodie espagnole* and
L'Heure espagnole to *Bolero* and the last songs, *Don Quichotte
à Dulcinée*, and also reappears in the piano music.

The world of children is evoked in *Ma Mère l'Oye* and
L'Enfant et les sortilèges with a tenderness and a nostalgia quite
absent from the nursery pieces of Schumann, Bizet, and
Mussorgsky. Like a child, Ravel identifies himself with the
creatures of his imagination in these pieces, as he does in the
animal portraits of *Histoires naturelles*, and his vision itself has
the objective, unspoiled quality of a child's. This is nowhere
clearer than in *L'Enfant et les sortilèges*. Here the non-human
impersonations of the household objects and garden-inhabitants
show an extraordinary resourcefulness and affection. But for the
child's first inkling of love, the little Princess who is 'the heart of
the rose', all Ravel can find is a passage of slightly sophisticated
Massenet. The very simplicity and naturalness of such a
situation defeated him. He was much more at home with the
comical or totally stylized 'love scenes' in *L'Heure espagnole*,
quite as much 'essays in a style' as moments in the drama; and
the still innocent, as it were, disembodied loves of Daphnis and
Chloe elicited music of a characteristically ethereal eroticism.

Behind his sophisticated and very carefully stylized exterior—
he was a great dandy all his life—Ravel in fact remained
emotionally immature. He would have admitted that for him
music resembled a child's box of toys which he enjoyed
arranging and rearranging in ingenious combinations—sardonic,
sentimental, or fantastic but never, except by occasional
implication, attempting any deeper significance. The instinct of
impersonation, or 'dressing up', extended to the art of composition itself, and he was one of the earliest composers, if not the
first, to model a composition of his own on a pre-existing

work—to sit down in front of a work by Mozart or Saint–Saëns 'as a painter sits down in front of a landscape'. The clearest example of this practice is the exquisite mock-Fauré slow movement of the G major Piano Concerto; but the keyboard works, which constitute Ravel's greatest claim to fame, show a similar indebtedness to Liszt.

His delight in orchestrating not only his own piano music but Mussorgsky's *Pictures at an Exhibition* shows this same fascination with what may be called the sartorial element in composition—designing, cutting, and ornamenting elaborate costumes for a fantastic and often purely imaginary pageant or masquerade. In *Le Tombeau de Couperin* and *Valses nobles et sentimentales* the dance-movements are as abstract as in any eighteenth-century keyboard suite, the masquerade non-existent; and yet the element of showmanship and sophistication is still strong, for these too are not direct utterances but 'music about music'. These close links between Ravel's music and that of the past—many pasts, in fact—won him a bad name with the younger generation of composers after the 1914–18 war. While Debussy's substitution of points of colour or decorative arabesques for continuous line was indisputably original, Ravel remained unequivocally linear; and he used the music of the past as a quarry, much as Stravinsky was to do, partly no doubt under his influence. When Ravel refused the Légion d'honneur, his detractors murmured, 'Yes, but all his music accepts it'—a classical instance of Establishment-smear. In time's perspective such considerations lose their force: J. S. Bach and Verdi were dismissed in much the same way by many of their contemporaries.

Where does Ravel's music stand today? Judged simply by the 'pleasure principle', very high indeed. Stravinsky's 'most perfect of Swiss clockmakers' referred to Ravel's syntactical and mechanical skill in composition, but 'most perfect of French cooks' would be equally true of his instrumental writing. Think of the cadenza in the Piano Concerto for Left Hand, 'Ondine', 'Une Barque sur l'océan', the dawn sequence in *Daphnis et Chloé* and even the opening of the Septet. These furnish aural pleasure of the most refined nature, never insistent or prolonged, sweet always offset with sour or bitter, perfectly textured and aerated, to produce an aesthetic euphoria very like the physical euphoria produced by a perfectly designed and cooked meal. Aural delight and aesthetic euphoria are by no means to be despised, but they have come to seem of secondary importance in even the secular music of Western Europe, which has hitherto remained for that very reason unique. It may well be that the days of that uniqueness are over, and it was certainly France that initiated the reaction against the conception of music as a quasi-religious, spiritual force. Ravel played his part in that reaction and in accepting his limitations he showed the instinct of a true artist, but the limitations were human and personal rather than aesthetic. I have not forgotten, after more than forty years, the sight of this diminutive man, exquisitely proportioned and economically built as one of his own creations, dressed in the magnificent crimson and white silk and black velvet of an Oxford Doctor of Music, for all the world like one of his own miniature fairy-tale figures all but lost in the silks and velvets of his creator's orchestra.

Albert Roussel[*]

Seven years younger than Debussy and six years older than Ravel, Albert Roussel had the brilliance—what, for want of a better term, we may call 'star quality'—of neither. His music has the qualities of good prose rather than poetry: logic, clarity, flexibility, well-judged variety of pace and of texture, and a general sobriety of colour, once the composer had paid his dues to the fashions of Impressionism and achieved his tardy maturity. Not that Roussel's music is in any sense prosaic; or that it bears any trace of that amateurish clumsiness that often betrays other late starters among the composers.

Roussel was born in Tourcoing, in French Flanders, one hundred years ago this month and was an orphan at eight, a circumstance that may well have played a part in determining the emotional reserve of his music. He entered the Navy and his seven years of naval service included a cruise in the Far East before, at the age of twenty-five, he finally decided to take up music professionally. Two years later, in 1896, he became Vincent d'Indy's pupil at the Schola Cantorum, where the severely classical, even academic, curriculum contrasted strongly

[*] The *Daily Telegraph* 12 April 1969.

with that of the Conservatoire, where Massenet taught composition and opera was still the central concern. Yet although
Roussel maintained his close connection with the Schola until
1914, first as a student and then as teacher (with Satie as one of
his pupils), his musical horizons soon became much wider than
d'Indy's; the solid musical foundations were used for super-
structures in which Roussel's personality became increasingly
apparent.

A *Divertissement* for piano and winds, dating from 1906,
often seems an uncanny anticipation of Stravinsky's *Petrushka*;
and when the Russian Ballet did in fact arrive in 1908, Roussel's
musical personality—like that of every other French composer
of his age—was strongly influenced by the revelation of a new
plasticity and a new exotic world of colours and forms. As early
as 1903, his settings of Henri de Régnier had revealed a strong
affinity with the ideals of Impressionism; and between 1912 and
1918 he completed three major works which exhibited his
mastery of the Impressionist orchestra and a strong but subtle
scholarly sympathy with India and its music, prompted by a
long visit to India, Ceylon, and Indo-China in 1909—*Le Festin
de l'araignée* (a ballet based on the insect-literature of Fabre and
Maeterlinck), the choral and orchestral *Evocations* and the
opera-ballet *Padmâvatî*.

In these works Roussel established himself as a peer and very
much a colleague of Florent Schmitt and Paul Dukas; and he
was to return much later to the ballet and the glittering
Impressionist orchestra in *Bacchus et Ariane* (1930) and *Aeneas*
(1935). In the meantime, however, the discovery of Stravinsky's
music prompted entirely new developments in his music and
enabled him to discover—at the age of fifty—the personality
which speaks unmistakably and with increasing vigour and
authority in the instrumental works starting in 1920 with *Pour
une fête de printemps*.

In the two orchestral suites (*Suite en Fa* and *Petite Suite*), the
Sinfonietta for Strings and the Third and Fourth Symphonies,
Roussel's musical language combines the fundamental seriousness and solid constructions of his Schola Cantorum origins
with a sardonic humour and a deep, though seldom explicit,
emotional commitment, both of which find expression in tartly
flavoured harmonies and a deliberate avoidance of easy elegance,

a preference for a certain clumsiness that recalls the rusticities of Flemish painting. It is music which clearly faces north, to the Channel coast (where in fact the composer had a house in which much of his composition was done) rather than to the Mediterranean.

His music had always been remarkable for its formidable rhythmic strength and variety, and this became increasingly marked—in the long ostinato of the *Sinfonietta's* slow movement, for instance, the self-renewing dance rhythms of the Third Symphony's Scherzo or the remorselessly heavy, thrusting gait of the Fourth Symphony's first movement. In the suites there is plenty of neo-classical, even archaic allusiveness; but the spirit of the music is always unmistakably modern, objective, disillusioned, or at least unindulgent. Moments of undisguised exultation, like the climax in the Fourth Symphony's slow movement, are the more rewarding for their extreme rarity, and Roussel makes no concessions whatever to the romantic reverie which, however disguised, appears persistently in Ravel's music.

The sound-picture of these works—texture, rhythm, harmony, though not actual material—and a certain impatient breeziness of musical temperament suggest an affinity with Walton, though Roussel's music is always plainly French. The classical French qualities, of which Roussel was the last great exponent in music, are shown most clearly in the chamber works, beginning with the Serenade for flute, harp and string trio (1925) and including a String Quartet and a Flute Trio, and in the forty-odd songs that cover his whole career as a composer, from 1903 to 1935.

Roussel's choice of texts reveals the same combination of fastidiousness and unsentimentality as his music. After the early settings of Henri de Régnier, the greater part are translations from the Chinese and the Greek, or single contemporary poems by Jean-Aubry, René Chalupt, and James Joyce, carefully chosen for their 'musicability'. If only a few have entered the general singer's repertory—'Le Bachelier de Salamanque' and 'Jazz dans la nuit' in particular—this is largely owing to Roussel's instinctive rejection of the obvious. Even when he seems to accept a traditional formula, such as the Spanish idiom in 'Coeur en péril', it is only to give it an unforgettably personal twist.

A musician's composer, then, not likely to achieve any securer place in the repertory than a very few works enjoy today? That might have been the verdict for some years after his death in 1937. Now, however, though the attention of professional musicians has turned for the most part to entirely different fields, either old or new, it may well be that the musical public would accept the comparative austerity and sophistication of Roussel's music and make that small step forward to acceptance which is all that is necessary for those who find no difficulty whatever in, say, Walton or Honegger.

George Frideric Handel[*]

Until some sixty years ago Handel was universally regarded by Englishmen as their great national composer, and even Elgar and Vaughan Williams have never really ousted him from his position of honorary laureate in perpetuity. His foreign birth, anomalous in a national hero, was tacitly disregarded during the nineteenth century, and our own day attaches little more importance to the circumstances than did the composer himself. The wars, persecutions, and emigrations of two generations have relaxed our sense of national individuality, and Handel's naturalization, an important sign of grace to our grandfathers, is no more than a bureaucratic fact to us. Yet his favoured position with English audiences is due in a large degree to the fact that generations of Englishmen have found in his music an idealized portrait of the national character as it was conceived in a less sophisticated age—frank, forthright, manly, and impatient of both frills and introspection. This conception of both Handel's music and the English character is certainly inadequate and in some respects wholly mistaken, but it persists as a strong sentimental memory which comes to the surface at moments of heightened national consciousness—a coronation, a royal funeral, or a great thanksgiving service. John Bull may be hardly more than a memory, but this memory is still potent when it is incarnate in a figure such as Sir Winston Churchill or finds musical expression in one of Handel's great choruses.

What sort of a man was Handel? When he settled in this country in 1712, at the age of twenty-seven, he was a cosmopolitan. He had spent four vitally important years in Italy,

[*] The *Daily Telegraph* 11 April 1959.

and spoke and wrote Italian and French in addition to his native German. The gross German accent with which he is supposed, in popular anecdotes, to have spoken English is almost certainly a fabrication. He was an opportunist, if by that we understand a man who takes a clear, objective view of a situation and cultivates those elements, both personal and artistic, which accord with his powers and inclinations and make success, in the widest sense, possible. What else should we ask of any artist, unless he is conscious of some overriding impulse to form rather than to follow the taste of his day, some 'mission' to explore and to innovate?

Handel worked, as the huge majority of artists have always worked, within the conventions of his age, which was not one of revolutionary discovery in the arts but of consolidation. He accepted aristocratic patronage as a matter of course, but for the last thirty years of his life he found himself catering for the new middle-class audiences which grew up as part of the economic expansion of his adopted country. If their character proved more moralistic and less frivolous or purely decorative than that of the court and the aristocracy, Handel found no difficulty in applying his gifts in a new way to suit a new public, and he moved from opera to oratorio without heart-searchings, professions of faith, or declarations of policy.

Musically he was always more Italian than German. English tradition, overwhelmingly coloured by awareness of *Messiah*, presents Handel as primarily a great master of oratorio. His contemporaries, however, thought of him as a dramatist, remembering his some forty operas, the odes, masques, and cantatas which display a wide range of human sympathy, a gift of powerful, direct expression, and an invention of orchestral colours and timbres which all belong by right in the theatre. There was, of course, a Johnsonian side to his musical character, but this has been exaggerated out of all proportion, so that we have risked forgetting the light-fingered, pastoral humorous composer of, say, *Acis and Galatea* on the one hand and the dramatic poet of *Solomon* and *Giulio Cesare* on the other.

His operas are constructed on the formal Italian pattern of the day which consisted of self-contained solo numbers linked by recitative, so that the drama moves stiffly from situation to situation, and the singer's task is to sum up, and only rarely develop, a dramatic mood. It was the singer's art that drew the public, and Handel's battles with these temperamental, vain, and jealous nightingales and the passionate factions which they aroused exhausted his energies during the best years of his life. The music that he poured into his operas has been neglected because the form itself is outmoded and the art which it demanded very largely lost. Revivals are often a bitter reminder of how much we have lost musically, but they never suggest the possibility of considering the form as anything but an historical curiosity.

When he turned to the oratorio, it was to escape from the financial anxieties and the ceaseless personal perplexities inseparable from the life of the operatic composer and the impresario—for Handel was both. The oratorios themselves are often more dramatic, in the modern sense, than the operas, because here the composer was less hampered by the conventions of the day and the demands of his performers. The recent performance of *Samson* on the stage of the Royal Opera House has demonstrated how few changes are needed to transport a Handelian oratorio back to the theatre where it would have belonged originally but for the unfortunate circumstances of the day. *Messiah* stands almost alone among Handel's works, the supreme example of a mixed style in which the operatic is very far from being forgotten (think only of 'Why do the heathen so furiously rage?' or 'He shall break them') but is blended with elements drawn from the German Passions and the English anthem. The music has not the note of intimate personal religious devotion that marks J. S. Bach's Passions and cantatas, but the broad drama and pathos, the rhetoric and the strong rational appeal of a great eighteenth-century homily or oration.

The cosmopolitan Handel found the public-spirited, reasonable, and decorous temper of Anglican Erastianism not only acceptable but the counterpart of one large element of his musical genius. Unlike Bach's, his music is essentially public, the music of a society rather than of a congregation; and even the instrumental works and keyboard suites suppose the mixed audience of the opera-house rather than Bach's passionate connoisseur or solitary, self-communing musician. Water-parties, firework displays, a national victory or a coronation—such events stimulated the sociable composer, who found no difficulty in writing his best music within the amateur's easy and immediate comprehension. This is no doubt part of the secret of his popularity with the English, whose dislike and distrust of the specialist, the fanatic, and the intellectual are as marked in their aesthetic as in their other sympathies.

It was characteristic that he had friends not only among the writers who frequented Burlington House and Cannons—Gay, Pope, Swift, and Arbuthnot—but in the financial world where he proved a shrewd investor. This 'all-round' character and the ability to move naturally and on equal terms with ordinary men and women of every class has been rare in the great composers, who have often been victims of their art, living in the enclosed world of their imagination. Not so Handel, and he has lost the sympathy and understanding of many great musicians—Liszt, Wagner, and Berlioz among them—by the resolute normality and extrovert character of his art. I fancy their disapprobation would have worried him little (it might have prompted one of his pithy, downright Anglo-German jokes) as long as he was assured of the affection and respect of the average musical man.

13

CRITICAL CONSIDERATIONS

Surprise*

The element of the unexpected plays a considerable part in the pleasure we derive from music, but it is not so simple a pleasure as one might imagine. In the first place there are two forms of musical surprise—the one purely physical (like the sudden fortissimo chord at the end of the slow and soft introduction to Weber's *Oberon* overture); and the other emotional or intellectual, or a mixture of the two (like the sudden change in the theme of the finale of Mozart's 'Haffner' Symphony on its final appearance). The second of these two forms of surprise naturally retains its character much longer than the first, which decreases rapidly and is always aimed by the composer at the groundlings.

There is, of course, some music in which the effect of physical shock is carefully avoided, and any form of surprise is kept at a very discreet level—Palestrina, for instance, and any music that respects the spirit of liturgical worship. At the other extreme lie the composers in whose music shock tactics are an essential part of the aesthetic scene—Gesualdo and Satie, and the whole race of Napoleonic and post-Napoleonic opera composers, chief among them Spontini and Meyerbeer whose object it was to 'send' (if that is a legitimate translation of *épater*) the new bourgeois musical public.

It was the French Revolution that raised the threshold of shock; and first Haydn and then Beethoven were influenced by the new use of the element of physical shock in music. But as naïve Viennese provincials, worlds removed from the Paris opera, they quite neglected the fashionable note of *terribilità*. They either made their surprises good-humoured jokes, designed

---

* The *Daily Telegraph* 27 October 1962.

to please rather than startle the audience, or else wholly transcended the physical shock by making it the occasion of a sudden new vision of the sublime. Beethoven was the supreme master of this, and there are countless examples in his music. Among many others, perhaps the most sublime is the sudden change of key, from A to F major, immediately before the Alla Marcia in the Finale of the Ninth Symphony.

It is these 'surprises' that retain their power after hundreds of hearings. Whereas we smile with Haydn's cosy jokes (or Beethoven's when he makes them) and at Rossini's *colpo di cannone*, or Weber's electric *sff*, Beethoven's sudden veerings or plunges of the spirit stir the listener today as profoundly as ever they did, and perhaps even more.

As the language of music has become more complex, and the possible range of dynamics greater, it has become increasingly difficult to achieve the effect of surprise. It has also become clear that the more consciously the composer seeks simply to shock or astound the listener, the more short-lived is his success. This has become very plain during the past fifty years, and can be explained by two facts. The first is the gradual disappearance of any established norm—melodic, harmonic, or rhythmic—departure from which has in the past constituted the surprise; and the second is the search for physically sensational innovation at all costs. This search has now defeated its own ends, so that it is no longer possible to surprise the experienced listener. Moreover even such well-motivated shock tactics as those used by Stravinsky in *Le Sacre du Printemps* or by Berg in *Wozzeck* are feeble in their effects compared with the great surprises of Beethoven's music. For the fact seems to be that surprise, in any but the most obvious and primitive sense of physical shock, can only be achieved when a language is still in its youth or maturity, not in its old age. Where are the great surprises in Wagner or Strauss, for instance? Both sometimes aim at surprising by sheer volume, both indulge in literary humour (Beckmesser and Till), but the purely musical unexpected is so rare in their works as to be negligible.

Stravinsky, in what the French aptly call his *pince sans rire* manner, included this kind of surprise in his general rejuvenation of the musical language. He was unique in seeing clearly, more than forty years ago, the diminishing returns of a purely external *terribilità* and changed his aesthetic accordingly, producing music whose exterior sobriety is in inverse proportion to its density and concentration of significance. Both Bartók and Schoenberg were to move in the same direction before they died; but in an age of technical discovery such as our own, the impulse to discover at all cost new techniques, or to make new use of the latest, is in many cases the strongest prompting in a young composer. At present, then, we find ourselves in the position of having an amply enlarged technical field with no new master to exploit its possibilities to the full. When he appears, he will certainly restore to music that element of the unexpected that it has for the moment lost.

Novelty*

Originality, which we now value so highly in all the arts, is not only a comparatively recent criterion of value but a very difficult concept to define. Novelty of form alone does not entitle a composer to claim originality or Spohr's 'Gesangsszene' Violin Concerto would constitute as good a claim as Beethoven's 'Choral' Symphony. Novelty of content, inasmuch as it is separable from form, is even less indicative, or John Cage would be a more original composer than J. S. Bach.

So what is this originality after which we hanker, and is this hankering a healthy appetite or a morbid craving? It is significant that it dates from the time of Beethoven, the first composer to emancipate music unequivocally from the church and the theatre and to raise the potential power of the art, unaided by anything extraneous, to hitherto unimagined heights. No earlier composer could have aimed at giving the whole essence of a drama in an overture, as Beethoven did in his *Leonore* No. 3, still less of writing a setting of the Mass which, like the *Missa Solemnis*, is itself a religious profession and experience, making the actual rite superfluous. Once music had been shown to possess this power of embodying a whole new vision of the world, any less ambitious objective came to seem too little for a self-respecting composer to claim, or rather for his admirers to claim for him.

'Heir of Beethoven' was a dangerous title very freely bandied

---

* The *Daily Telegraph* 13 March 1965.

about Northern and Central Europe for the rest of the nineteenth century. It was wished at various times on Brahms (who found the weight of the crown crippling), Liszt, Wagner (who, like Napoleon, placed the crown on his own head), and finally, and most improbably, on Sibelius, whom it perhaps shocked into the long silence that followed the Seventh Symphony. The Latin and Slavonic countries remained mercifully free from this neurosis and their music flourished accordingly during this period, while Germany gradually developed the gigantism or acromegaly which proved the penultimate stage in her decline as a musical great power.

Yet when Wagner's music became universally synonymous with avant-garde art in every country—about the 1880s—the composer found himself automatically elevated to an unnatural position, half priest and half prophet, and his works were regarded as pronouncements of which originality was an essential characteristic. (How cynically an earlier generation had regarded such pretentions can be seen from Bizet's malicious remark about Gounod—'Art is for him a kind of priesthood. He told me so.') It was in protest against such inflation that Satie embarked on his sly, pseudo-primitive miniatures and Stravinsky and his neo-classical followers resurrected the ideal, and the music, of the minor eighteenth-century craftsmen-composers.

Yet the idea persisted that each new work must be unique of its kind, no longer a spiritual but a stylistic revelation or an enlargement never previously attempted of music's language. If Hindemith moved furthest away from this quest from originality and came nearest to creating a new mid-twentieth-century lingua franca, drab but serviceable, the New Viennese School remained most deeply committed to the nineteenth-century romantic ideal. Schoenberg summed up this highly-strung, perfectionist attitude to art when he said that what was worth saying was only that which had not been said before. This suggests at once an art which has to create, rather than to respond to a demand; and here, I believe, we have at least one root cause of the cult of originality—namely, that goods for which there exists little or no demand must be at least apparently unique of their kind if they are to attract a market.

We may reasonably hope that, as the musical public becomes

better acquainted with contemporary music and gradually develops a natural taste for it, the composer will feel relieved of the feeling that he must at all costs innovate. After all, this is quite a new idea. Handel happily quoted large portions of one work in another and Bach delighted to rewrite Vivaldi, as Stravinsky has rewritten Bach. Composers confident of public interest in their music—such as Britten and Shostakovich—are already able to say, quietly and in their own voice, things which they have said in slightly different ways before in other works. This is not to say that Britten has lost the exploratory instinct: *Curlew River* makes it clear that he has not. But he is no longer under the compulsion to shock, startle, alienate, or mesmerize his listeners. He can once again regard them, as listeners have hardly been regarded since the eighteenth century, as friends and, in a sense, patrons—the natural consumers of the goods he produces instead of enemies, cretins, and bourgeois fools incapable of appreciating his inspirations.

In fact, the heroic age of the musical revolution may be considered over, and those who continue the firing can generally be dismissed as left-wing deviationists or children who have rifled an ammunition dump. True originality will never lack admirers; but we are learning to look for it in less obvious forms—in the quality of the musicial ideas and the individuality of their treatment rather than in eccentricities of language, subject-matter, or presentation.

Values that pass[*]

One of the arguments recently put forward for retaining at least rudimentary instruction in the doctrines of Christianity in schools was the need to preserve an understanding of European, especially medieval, art which is so largely based on Christian theology, morals, history, and literature. A Jesse window showing the Old Testament genealogy of Christ, or the sculptures of the Chartres porches, must in fact be rapidly becoming as abstruse to the average European as the labours of Hercules or the loves of Jupiter, matters of erudition rather than of everyday knowledge.

Nor are we faced simply with questions of iconography, the

[*] The *Daily Telegraph* 11 July 1970.

visual or literary references to historical or mythological characters. The whole ethos of European art, founded in moral and religious presuppositions that needed no explicit statement, has been slowly changing for at least two hundred years; and that process has recently been so accelerated that we now, for the first time, find ourselves in a world where no scale of values can be confidently predicted in approaching any work of art. We have passed through the 'transvaluation of all values' foreseen by Nietzsche and entered a no-man's land where the very word 'value' has a faintly ironical ring. The whole future of our civilization depends on how this vacuum will be filled, for filled it will certainly be.

Meanwhile strange anomalies begin to appear. How long will it be, I found myself wondering at Glyndebourne the other night, before an audience will need a footnote explaining the enormity of Tatiana's offence in declaring her love to Onegin without any preliminary suggestion that it was returned? Does it not already begin to seem, as a dramatic motive, almost as unreal as Antigone's concern with burying her brother's body, a concern so overmastering that she was ready to die for it? Only a deliberate act of the informed imagination can make us understand and sympathize with Antigone today. But is the ideal of woman's chastity, against which Tatiana's letter was felt to be an unmistakable offence, likely to survive in our society even as an imaginative element in a work of art? If not, this would explain the difficulty in finding young singers able to give a natural and convincing performance of Onegin's role—able, that is, to convey the exact shade of cool condescension that Tatiana's action entitled him to adopt, the infinitely wounding magnanimity with which he consented to overlook her shamelessness.

At Glyndebourne even the quarrel between the two friends seemed shadowy and unreal. We have gone so far towards losing the proprietary male attitude towards women, at least unmarried women, that Onegin's mild flirtation with Lensky's favourite Olga could not conceivably lead to anything more than a rough passage of words between them. Duels of honour, though even more completely outmoded than the ideal of feminine chastity with which they were often closely involved, have at least the advantage of action; and since no kind of fighting has, unfortunately, vanished from our emotional consciousness, a stage fight is, dramatically speaking, its own justification.

Such puzzles as these are by no means confined to opera. All lovers of Renaissance music must have been struck by the attraction exercised by Petrarch's sonnets over the composers of the day, and wondered what significance these apparently frigid, conundrum-like sweetmeats concealed. (Perhaps it is not difficult to imagine the Heine/Schumann *Dichterliebe* presenting a similar enigma to our not so remote descendants.) Petrarch continued to interest a few composers long after the Renaissance, including Liszt and Schoenberg; and it was listening to some performances of Liszt's 'Sonetto No. 104' from the *Années de Pèlerinage* that I began to understand Petrarch's special appeal.

The chief characteristic of Petrarch's poems is the combination of an extremely musical, mellifluous language with a strict, tortuous syntax that makes a directly intellectual demand on the reader; and this is a particularly happy image of frustrated love, i.e. the enforced submission of sexual desire to limitations dictated by principle, moral or religious. This frustration of the sexual instinct has been taken so much for granted as a principle of society, and one of the prime sources of European art, that I was amazed to find that the erotic mood of Liszt's piece entirely eluded the two young players concerned, who seemed aware neither of the passion involved nor of the fact that it draws its strength from being compressed or frustrated, like a banked fire or a dammed stream. They were doubtless sexually experienced, but in the manner of *Turangalîla* rather than *Tristan*; and this for the simple reason that in their psycho-physical lives they had encountered no real barriers, nothing capable of building up the pressure and tension that transmute physical desire into 'romantic' or ideal love, as heat and compression convert water into steam.

As the conformation of our emotional world changes, so must our understanding of the arts of the past; and the effect may go deeper even than the erotic, to touch what have hitherto been accepted as the springs of human feeling. I have heard a famous Russian sing the soprano solo part in Beethoven's *Missa Solemnis* without once suggesting that she knew the meaning of the word 'supplication', the cry for mercy and forgiveness that sounds through the Kyrie, the Agnus Dei, and much of the Gloria. This, it may be, is the truly modern human being, consciously self-sufficient and subject only to chance political or economic necessity; certainly not a creature, as Beethoven imagined, by his very nature dependent on a creator. Such a feeling was outside this singer's emotional range. Perhaps, indeed, it is not enough for an artist to experience this feeling in his imagination only; there must be some contact with it in his own emotional experience if he is to communicate it to an audience. In a similar way no Western singer I have ever heard came near to catching the note of unquestioned autocratic demandingness that sounded in the voices of Chaliapin and Christoff when they sang Tsar Boris. I recognized it as genuine at once, but as though from far away, something imagined but never in fact experienced.

Humour[*]

The disappearance of the last dance movement from the symphony was a minor sign of the times. The minuet was essentially a court dance, and even Haydn's minuets have often lost their original character. But it is a strange fact that the replacement of the minuet by a movement entitled 'scherzo' should have coincided with the full maturing of the symphony as the most serious and comprehensive of all abstract musical forms. The Italian word was borrowed directly from German, just as German borrowed from Italian that other word for joke, *Spass*, which Mozart used for his K. 522. Music could show plenty of examples of wit before the scherzo, but musical wit demands (like verbal wit) some knowledge of the language in which it is couched and so presupposes an educated audience. An extreme example is the enharmonic pun, or play on the double meaning of a chord according to its context, most familiar in Rimsky-Korsakov's *Scheherazade*. But the carefully pointed dialogue of Mozart's piano concertos or a movement like the Finale of the 'Haffner' Symphony, where the listener's expectation is suddenly deceived, are examples of the playfulness of the creative spirit within its own field of reference, like the wit

[*] The *Daily Telegraph* 17 April 1976.

of Shakespeare or, indeed, Sheridan among Mozart's contemporaries.

Humour is a different matter, and jokes belong to the world of humour rather than to that of wit. It is essentially a plebeian world, and it was not by chance that it was through 'low life' interludes in tragic works that comic opera came into existence; or that bourgeois and peasants re-enter the world of serious opera through a door leading to the department of comedy. The distinction between wit and humour is not clear cut, but can often be determined in practice by observing the listener's reaction. Wit prompts a smile, that marks only the listener's face, while humour prompts a laugh that may shake the whole frame.

To Aristotle, whose 'gentleman' provided in the last resort the ideal of aristocratic behaviour, laughter was for this very reason unseemly, a momentary relaxation of the self-control that distinguished the gentleman from the boor. Nothing of Mozart's, except, significantly his *Musikalischer Spass*, prompts laughter as do Haydn's simple jokes and Beethoven's deliberate fumblings with wrong notes, the horn's 'wrong' entry in the First Movement of the 'Eroica' or the timpani's interruption of the Scherzo of the Ninth Symphony. Haydn's humour belongs to the village world from which he sprang and its court clothes deceive no one, but merely ensure an entry into polite society. Beethoven's humour ranges from the primitive banana skin of the horn in the 'Eroica', through the personal eccentricity of the canons on his friends' names, to the Homeric mirth of the Scherzo of the last quartet, Op. 135, where the fifty-odd repetitions of a single phrase in the trio play with obsession in a way that may not be straightforwardly comic but is certainly not witty.

Humour, it may be thought, is not really at ease in music as such. The sophisticated buffoonery of Chabrier was soon followed by the cutting sarcasm of Prokofiev and the jokes of Satie, to end with Cage's No-jokes and Hoffnung's slapstick. Only the opera provides worthy parallels to the works of the great comic writers, and these are few enough. Italian and French comic opera in the eighteenth century is mildly witty; and that wit which was still essentially rooted in the aristocratic mockery of the unprivileged by the comparatively privileged, was lent a double edge and a unique musical quality by Mozart. In Rossini, wit is often diluted by humour, sometimes primitive in the early works but sophisticated enough in *Le Comte Ory* and the final *senilia*.

The two great comic masterpieces of the century, after *Il Barbiere di Siviglia*, are of course *Die Meistersinger* and *Falstaff*. In these there appears uniquely that broad human sympathy founded on, when all is said and done, a deeply optimistic view of human nature, liberally qualified no doubt but unmistakable since its disappearance and only possible in a society which still felt fundamentally secure however threatened. The comedy of *Die Meistersinger* is darkened by the malicious portrait of Beckmesser. None of Molière's or Dickens's characters are humiliated with the plain delight that Wagner brings to Beckmesser's downfall, in which it is impossible not to feel a personal element. By comparison, Falstaff's discomfiture is good-humoured and Verdi's final fugue is significantly led by Falstaff himself with the conclusion that 'all the world's a joke'. (Boito uses the word *burla*, from which 'burlesque' is a derivative; and there is perhaps an unconscious reference to the Italian proverb—*chi burla si confessa*, i.e. you pretend to be joking but we know that it is true.) The Boito/Verdi *Falstaff* is quite without the puzzling overtones of Shakespeare's complete portrait and is indeed the Falstaff of *The Merry Wives of Windsor* only. In spite of the dark elements in Verdi's own character, the comic world of his last opera is quite unclouded.

Satire and farce, which are all that is left of the comic spirit in our disoriented society, found classical expression in Offenbach, the all-too-shrewd court-fool of the French Second Empire. Not for nothing were his works appreciated by Nietzsche, the first of the major prophets of doom, and Bismarck, the engineer of the downfall of the particular society which Offenbach satirized; and most significantly of all by Napoleon III himself. Is it still too late to suggest Ionesco as a librettist to Cage?

Competitions[*]

Competition—the instinct to do better than one's neighbour—is natural in all but a very small minority of the human race. This

[*] The *Daily Telegraph* 20 December 1980.

instinct may be encouraged or discouraged, reduced to a minimum or exaggerated to become the mainspring of life; but it cannot be eradicated, and it is essential to the achievement of excellence in almost any field. Utopian, or simply high-minded, thinkers have always been there to remind us that one man's success is another man's failure, that failure in one race does not mean failure in all, and that success is itself an ambiguous word except within very narrow limits. In a school examination, in a race or a competition for any kind of prize, success and failure are presented as unambiguous; but successful careers are by no means confined to prize-winners and youthful winners of many prizes often remain in the background for the rest of their lives.

In musical competitions the problem is still further complicated by the fact that even the terms of the competition are hard to define. Does A play Bach better than B plays Beethoven is a question that cannot logically be put, let alone answered. Furthermore 'better', in such a context, is itself ambiguous, since technical control of an instrument and understanding of music are two different faculties combined in quite different proportions in the same performers, even when they have reached maturity. No doubt it is an awareness of this, honourable in origin but sentimental in practice, that prompts Anglo-Saxon juries in particular to multiply—and even divide—prizes and to add 'honourable mentions' and 'runners-up' to their results in order to soften the blow of failure by words of encouragement and promises for the future. A proud competitor may well resent being coddled in this manner, which in any case contradicts the very nature of competing for a prize—something to be won or not won. The hesitation in awarding a prize and thus declaring that one competitor literally excels others comes, I believe, from two sources.

First, there is the realization that artistic performances cannot be judged like 100-metre races, with one runner indisputably first. But there is also today a reluctance to anticipate that 'natural selection' which will eventually weed out the weak from among the strong, a reluctance even to voice a clear distinction of individual quality. Nature, on the other hand, is a great élitist who awards very few consolation prizes; and it might well be kinder to inure the young artist to short-term failure rather than seek to palliate the unpleasant truth.

Some years ago, Claudio Arrau reviewed the question in a magazine article in which he admitted most of the above objections but still found competitions justified by the publicity which they give to young talent and by their psychological effect on competitors. He did not mention the mental and emotional strain involved or the impact of failure, which must still affect the great majority of all competitors. He spoke only of what winning a competition should mean for the winner. Victory, he says, is only 'a mandate for further growth'. Although there are unavoidably 'elements of the circus' in anything competitive, the winner must immediately set these things aside and dedicate himself to a life of 'hard work, study, thought, reflection and, quite importantly, humility'. He must do this moreover with the eyes of the world on him and concert agents badgering him to exploit his success and embark at once on a career. In fact 'from success as a competitor to success as a musician is the longest, loneliest road' and not one travelled by all prize-winners. Indeed it requires in Arrau's words, 'the greatest courage and utmost dedication', that is to say, great strength of character.

If then competitions only really benefit those who win them—which cannot include more than those placed among the first three—and if winning presents such a formidable moral problem to the winner, do they serve any substantial purpose? Negatively their effect can certainly be disastrous. In a field where real progress is of its very nature slow, since it involves the whole personality, competitions demand an artificial acceleration, a process comparable to inducing the birth of a child. The legitimate element of competition in a young musician's mentality is unnaturally stimulated, and premature publicity—even for the losers—militates against Arrau's ideal of study, thought, reflection, and humility. Almost inevitably 'success' becomes identified with prize-winning rather than with growth as a musician, and this can easily result in a race of performers whose often superb command of their instruments *in extenso* is not matched by the quality of their musicianship *in intenso*. Is it, perhaps, better neither quite to lose nor quite to win in a competition, as is now possible? Or would music rather benefit if competitions were reduced, or even abolished?

Projection in performance[*]

It is a commonplace in the theatre to speak of an actor being 'inside' his role, that is to say identifying himself with the psychological states of the character that he represents. The actual 'projection' of the part, though essential, is quite a secondary matter. The same is true of musical performance, but with a slight yet important distinction due to the nature of music. This—though to most people a vaguer form of communication—should have for the musician, as Mendelssohn once observed, a precision of significance greater rather than less than that of words. To put it bluntly, it is much easier for the musical performer than for the actor to persuade his audience that he is inside his role when in fact he is not. Technical skill, charm of manner, and a little discreet mimicry—showmanship, in fact— will enable a forceful and determined personality to persuade ninety per cent of most audiences that they are encountering a great artist. For what the great majority of music-lovers still want is to be dazzled by technical display and to be taken out of themselves by contact with an emotional world more highly-coloured than their own.

This is a real hunger; and like all really hungry people, the large public is not fastidious. They still like pianists who play very loud and very fast, singers who sing very loud and very high, and violinists who combine an ethereal top register and a schmaltzy G string with breathtakingly fast passagework. Now a great many of the truly great artists can satisfy all these demands, as it were incidentally, and also give much more—the music itself, in fact, as well as its more obvious features. But there have always been, and still are, many musical performers who, if they were actors, would very soon be seen to be hamming their way through a part, producing, that is to say, the stock reaction to each situation. Such players have always formed the majority of those who appear before the public and include today also quite a number of famous names. They very seldom possess any range of soft piano tone, though they can make a great impression from *ff* to *ffff*; and they normally travel in the middle lane between mezzo forte and forte. This poverty

[*] The *Daily Telegraph* 3 May 1975.

of dynamic range is matched only by that of their range of true espressivo. I say 'true' because it would be a mistake to say that their playing lacks light and shade, that it is rhythmically wooden or obviously insensitive. On the other hand, each marking of the composer is often observed as though it were a stage direction, always evoking the same reaction whatever the context, just as an organ-stop produces a mechanically identical sound every time that it is drawn.

This 'simulated emotion' is, I believe, the greatest single enemy of the performing musician; and the first to be deceived by it is the performer himself. The quick instinctive sympathy and wide-ranging sensibility of the true artist are quite different from the manual dexterity or oral agility and the extrovert, even exhibitionist, character of the natural performer; and it is only the combination of the two that makes a great performing artist. The artist of sensibility who lacks the performer's gifts is automatically debarred from appearing in public; but the natural performer will often make a career despite a very meagre endowment, if he picks up the conventions of musical expression and imitates them well.

Since an ability to imitate is an essential element in the performer's temperament, it is almost certain that he will unconsciously pick up these conventions at a very early age. Thus we are faced with a dilemma of the 'can't lead, won't follow' variety: those with a natural gift for performance, i.e. communication, too often have little of interest to communicate; while those who have most to communicate too often lack the psychological and perhaps the pushy, physical—that is, technical—means of communication.

Those in the first class used formerly to rely on making a spectacle of their own personalities (a few still do) and such players were admired for their 'temperament'. (English education at all levels, with its horror of 'show-offs', has been an almost insufferable handicap in this field.) A commoner alternative today for those with meagre or shallow musical personalities is the combination of a showy technique (fast tempos and high decibel-count) with a fluent use of the accepted conventions of emotional expressiveness.

This includes a certain amount of miming, though much less than in the romantic period when these conventions were first elaborated, and much over-emphasis of all kinds. Times change, however, and it may well be that exaggerated tempo rubato, and even the non-synchronising of the pianist's two hands may reappear in the concert hall. Anyone who wishes to savour the effect of such methods, presumably in their more austere form, has only to listen to the recording of Busoni playing Liszt's *Rigoletto* paraphrase. Meanwhile, all those who serve on musical juries will be familiar with the endogenous disease of 'stock response' or simulated emotion. In fact, a young player who is absolutely free of it is so rare that he can be fairly sure of favourable notice at the least and probably an award.

Italy*

Kennst du das Land, wo die Citronen blüh'n? Goethe's Mignon sums up the nostalgia, the sense of almost physical longing, that Italy has inspired in generations of Northern European artists, not least musicians. Italy's own contributions to the arts provide only part of the spell, and it is rather the serene harmony of an idyllic landscape and an apparently carefree and uninhibited people with the monuments of their architecture, painting, or music that has drawn the Northern artist from what Ruskin called his 'foggy taverns'. Goethe's accounts of his crossing into Italy and the sense of stepping into a new, though infinitely older world in which all the senses are heightened, has plenty of parallels. It appears as an intoxication in Meyerbeer's letters, as the precipitant of a moral crisis in the sensitive young Mendelssohn ('Much that was thought unchangeable and permanent has been swept away in a couple of days') and eventually provoked a violent reaction of nostalgia in all the great nineteenth-century Russians, Glinka and Tchaikovsky as well as Gogol.

Almost all these artists paid their homage to Italy in a number of works, and cherished Italian memories, probably indistinguishable from memories of their youth. We owe Sibelius's comparatively sunny Second Symphony and Violin Concerto to his stay at Rapallo ('The Mediterranean in storm! All the small birds are here. They shoot at them. But the birds sing and wait for the spring. Finland! Finland! Finland!!!'). The flower-

* The *Daily Telegraph* 7 October 1967.

maidens in *Parsifal*, though imagined at Ravello, seem only to confirm our impression of Wagner's bone-German-ness. Berlioz's case is slightly different. The French cannot resist a certain feeling of superiority to Italy, as to a faintly ludicrous and shockingly unintellectual sister (a relationship whose other side is reflected in the Italian Press's bitter-sweet references to *nostra cara sorella latina*) and Berlioz was no exception. His insistence on pilgrims and bandits in *Harold in Italy*, his ignoring of all but the obviously picturesque sides of Italian life might be compared to those of an American composer visiting this country who divided his four movements between Trooping the Colour, Last Night at the Proms, Westmorland Sheepdog Trials and the Beatles. These would all be legitimate inspirations, but would not reveal any profound feeling for England. The French have, of course, their own equivalent to Italy in Provence, a different but similar region; and composers (Bizet, Debussy, Ravel) have found more stimulus in the rougher, tougher, less sensually appealing life, landscape, and music of Spain. Here again, however, Spanish composers have sensed a note of condescension in the 'picture postcard Spain' of *Carmen*.

What kind of music has Italy prompted in most of her passionate Northern admirers? I think we can discount J. S. Bach's Vivaldi-rewritings as based entirely on musical grounds and quite uninfluenced by any ideal picture of the climate, landscape, or atmosphere of Italy. Northern composers who visited, or even settled, in Italy before 1800 did so for strictly professional reasons (the Netherlanders in Venice and Rome, J. C. Bach and Gluck in Milan, and Hasse in Naples) and if their music took on a noticeably Italian colour, this was for strictly economic rather than sentimental reasons. The boy Mozart plainly enjoyed his Italian sprees, without any great awareness of cultural overtones, though he was stimulated by the experience of Italian opera in its own setting. If Italy makes itself felt in the quartets K. 155–60, it is exclusively because of the opera.

An awareness of the atmosphere and character of Italy, or any other country, really entered music as part of the romantic mentality. Whereas the Italian moments in Schubert's symphonies are simply reminiscences of Rossini, and purely musical, with the second volume of Liszt's *Années de Pèlerinage* we have a real attempt to reproduce in music the specifically Italian qualities found in Dante and Petrarch, as well as the more 'touristic' reactions in the supplement. (This concern with Italian literature is very rare. Most foreigners have felt with Browning that 'the Italians are poetry, don't and can't make poetry'.) In the third volume Liszt takes the gardens of the Villa d'Este to typify one whole side of Italian landscape and explores its musical possibilities in not at all touristic terms. But the sensitive, chameleon-like Tchaikovsky was perhaps the first composer to communicate in purely musical terms the feel of Italian life, not in terms of art or landscape but in the popular melodies, street sounds, bugle-calls, and so forth of the *Italian Capriccio* (1880).

Only a dozen years later there appeared at the other extreme a unique recreation of a purely imaginary Italy in the first volume of Hugo Wolf's *Italienisches Liederbuch*, following the *Italian Serenade* for string quartet (1887) and itself to be followed by a second volume of songs (1896). Although Wolf's native Styria is within a stone's throw of Italy by today's reckoning, he never went there, and the texts which he chose were Paul Heyse's German workings of Italian originals. Wolf's songs are certainly more Austrian than Italian (perhaps comparable in this to Britten's very English settings of Rimbaud and Michelangelo) particularly in the intense interiority of the serious songs and the self-consciousness of the witty. In the *Italian Serenade*, on the other hand, Wolf achieved an ideal expression of that light- and open-heartedness, that innocent sensuality without a trace of acridity and that innate, instinctive cantabile character which Northerners, sometimes rather too naïvely no doubt, find in Italian life and art. In fact, these Italian-inspired works of Wolf's might be regarded as the obverse, or even the ransom, of the Viennese addiction to Puccini.

No English composer until Walton has settled in Italy, though Elgar paid his tribute with *In the South*, written at Alassio but in a vein apparently predetermined by the composer's reading of Virgil and Byron and the work of an observer of the Italian scene rather than a participator. Certainly nobody could embody less the Italian idea of *inglese italianizzato, diavolo incarnato* (the Englishman turned Italian, a devil incarnate) than Sir William Walton. If his *Troilus and Cressida* was written with an eye to La Scala, this was in the eighteenth-century tradition of practical rather than emotional considerations and it would be hard to find any Italian influences, realistic or ideal, in the rest of the music. Henze is the obvious example today of a Northern artist who not only settled in Italy but, for a time at least, Italianized his art. Three hundred and fifty years earlier Schütz had done the same without sacrificing his own essential character. If Henze is more of a chameleon by nature, ours is also a chameleon age, and Italian life can still be an education for any human being, and how much more if he is a composer.

PART IV   SONGS AND SINGERS

14

SCHUMANN'S SONGS[*]

'Do you, I wonder, feel as I do?' wrote Schumann to Hermann Hirschbach in June 1839. 'All my life I have thought vocal music inferior to instrumental and have never considered it to be great art.' Only eight months later, in February 1840, he wrote of 'composing nothing but songs' and in a letter to Clara: 'Since yesterday morning I have written nearly twenty-seven pages of music—something new, of which I will only say that I laughed and wept for joy as I wrote. . . . Oh! Clara, what a joy it is to write for the voice, a joy I have lacked too long.' That is the typical Schumann, a theorist whose theories are always apt to be contradicted by the mood of the moment, emotional and unpredictable and glorying in the quick succession of extremes of feeling, the sun chasing the clouds, laughter and tears at one and the same moment, creating in his music the rainbow effects which he so much admired in the prose of his beloved Jean Paul Richter. Up to thirty the piano had been virtually his only confidant—for that is exactly what Schumann's piano music represents, confidences—but with the final blooming of his love for Clara Wieck he felt the need of a still more personal and intimate form of expression. Of Schumann's two hundred-odd songs very nearly half were written in the year of his marriage (1840) and that half contains almost all the best songs he was ever to write.

In spite of his letter to Hirschbach, Schumann had already tried his hand at song-writing before 1840. His Op. 11 originally consisted of eleven songs dedicated to his three sisters-in-law. Three of these were published by Brahms in the supplementary volume of the Collected Edition; six by Geiringer in 1933, and one—a setting of Goethe's 'Der Fischer'—as a

---

[*] G. Abraham, ed: *Schumann: A Symposium* (London, 1952), pp. 98–137.

supplement to the *Zeitschrift für Musik*, also in 1933.[1] But Schumann was not satisfied with them as songs and used three of them in his piano-works of the 1830s, 'An Anna II' (composed 31 July 1828) appeared as the Aria in the F sharp minor Piano Sonata; 'Im Herbste', as the Andantino of the G minor Piano Sonata; 'Der Hirtenknabe' (composed August 1828) in the Intermezzo Op. 4 No. 4. In fact the new lyrical impulse which Schumann brought to his piano-writing made the distinction between instrumental and vocal music more blurred than it had ever been before; and made him, too, the ideal link between Schubert and the next generation of song-writers, for whom voice and instrument were of equal importance. His literary taste and affinities gave him a feeling for prosody and a sensitiveness to the atmosphere of a poem such as no previous song-writer had ever had. A glance at the list of poets set by the mature Schumann gives an idea of his literary taste. Heine easily heads the list, with 42 poems. Then come Rückert (27), Goethe (19), Eichendorff (16), Justinus Kerner (14), Chamisso (11), Lenau (10), Burns (9), Geibel, Mary, Queen of Scots, and Hans Andersen (5 each), Mörike and Hoffmann von Fallersleben (4 each), Schiller (3), and Tom Moore (2). Of nonentities he hardly ever set more than a single poem, except in the case of the young poetess Elisabeth Kulmann, whose romantic life and early death at the age of seventeen led Schumann into mistaking her for a genius. 'Wilfried von der Neun' (whose real name was F. W. T. Schöpff) was no more than a talented amateur, but his poems at least provided Schumann with the larger and vaguer background that he needed in 1850, and they are certainly superior to the pretty platitudes of Elisabeth Kulmann.

Heine was the ideal poet for Schumann not only because a certain spiritual affinity existed between them, showing itself in the deliberate cultivation of sharply contrasted emotional moods within a single lyric; but also because of the conciseness and point of his style. Schumann, with his admiration for Jean Paul, was quite happy dreaming his way through the most circumstantial and flowery writing, pursuing the most far-fetched metaphors to their logical and ludicrous conclusions, savouring the extremes of comedy and sensibility on the same page and quite oblivious of the absence of plot or formal

---

[1] The eleventh, Jacobi's 'Klage', was apparently never finished.

arrangement.[2] Heine's poetry served him as an unconscious discipline, curbing his natural tendency to divagation and forcing him to come to the point, to condense his emotions to their sweetest and their bitterest. When, in July 1828, Schumann sent his first songs to Gottlob Wiedebein, the Brunswick Kapellmeister, the advice he received was to 'look to truth above all. Truth of melody, of harmony, of expression—in a word, poetic truth.' It was really Heine who enabled Schumann to follow that advice.

Thirty-seven of the forty-two Heine settings were composed in the year 1840. Heine's ballad 'Belsatzar' was one of the first poems he set, on 7 February of that year (although it only appeared six years later, as his Op. 57) while the first of the opus numbers consisting of songs was the Op. 24, the *Liederkreis* of Heine poems, dedicated to Pauline Viardot. His Opp. 25, 27, 29, 30, 31, 34, 35, 36, 37, 39, 40, 42, 43, 45, 48, 49, 53, 57 were all written the same year and a study of these songs alone would give a complete idea of Schumann as a song-writer. In the next twelve years he never wrote anything better and only occasionally anything nearly as good as appeared in this sudden enormous spate; and as he grew older his literary instinct seems to have faltered. How important this was to the quality of his music he was himself perfectly aware. 'Parallel to the development of poetry, the Franz Schubert epoch has been followed by a new one which has utilized the improvements of the accompanying instrument, the piano', he wrote in his *Observations on Composers and Composition*. 'The voice alone cannot reproduce everything or produce every effect; together with the expression of the whole, the finer details of the poem should also be emphasized. All is well as long as the vocal line is not sacrificed.' It is not surprising to find the piano playing a more prominent part in the songs of 1840 than in the later ones,

---

[2] In case the reader may tire of reading the name of Jean Paul repeatedly or may suspect that modern denigration of his style is unjust, I choose at random one of his high-flown metaphorical passages. This is taken from the *Flegeljahre* and Vult is speaking. 'O reiner, starker Freund, die Poesie ist ja doch ein Paar Schlittschuh, womit man auf dem glatten, reinen, krystallnen Boden des Ideals leicht fliegt, aber miserabel forthumpelt auf gemeiner Gasse.' ('My pure, strong friend, poetry when all is said and done is a pair of skates, on which one can skim lightly over the smooth, clean crystal surface of the Ideal, while in the ordinary street they make one's progress a wretched hobbling.')

for Schumann was still first and foremost a pianist at that point. What is more surprising is that no single one of these early songs is a piano solo with obbligato voice. (In fact it is possible to detect, on internal evidence alone, the authorship of No. 2 of Op. 37, one of the three songs contributed by Clara to the twelve settings of poems from Rückert's *Liebesfrühling*. 'I have no talent at all for composition,' sighed poor Clara and certainly 'Er ist gekommen' is simply a piano solo with a voice part added, just the song one would expect a piano virtuoso to write.) The preludes and postludes, generally considered to be typical of Schumann's song-writing as a whole, most frequently occur in the 1840 songs and are nowhere so prominent as in the *Dichterliebe* (Op. 48) and *Frauenliebe und -leben* (Op. 42) series. Another feature which is strongly characteristic of the 1840 songs and hardly appears afterwards is the turn. This appears in 'Widmung', 'Lied der Suleika' (Ex. 1), and 'Aus den

EX.1

Wie mit in-nig-stem Be - ha - gen,   Lied, emp-find! ich dei - nen Sinn!

östlichen Rosen' of Op. 25 and, 'Er, der Herrlichste von allen' and 'Helft mir, ihr Schwestern' of Op. 42; but on the rare occasions when Schumann uses it in his later songs ('Himmel und Erde', Op. 96 No. 5, for example) it strikes a false note and seems out of style. Whether the turn is to be considered as fundamentally a pianistic trait—a habit of the fingers in a composer much given to improvising—or to be related to Schumann's unconscious imitation of the operatic style, it remains a distinct mark of the year 1840.

The fingers obviously suggested another trait which is common throughout the whole range of Schumann's songs but particularly so in those of 1840. This is the anticipation of the voice by the piano or vice versa, such as we find in 'Es treibt mich hin' (Op. 24), 'Intermezzo' (Op. 39) and 'Aus den hebräischen Gesängen' (Op. 25) (Ex. 2) in one form and in 'Stille Liebe' (Op. 35) in the other. Occasionally this device of syncopation, which became almost a mannerism in Schumann's piano works, is used with programmatic effect, as in 'Lieb Liebchen' (Op. 24) where it represents the beating of the lover's heart and gives rise to the subtly dramatic ending of each quatrain, where the words 'Totensarg' (coffin) and 'schlafen kann' (can sleep) are left hanging in the air by an accompaniment which has come to an end a bar earlier. These illustrations or programmatic devices are rare in Schumann's songs, but when he attempts them they are almost always carried out with the finest musical sensitiveness and never exaggerated. The fall of the rose petals in 'Der Hidalgo' (Op. 30) and the rippling of the waves in 'Aufträge' (Op. 77) are effective because of their simplicity and their musical unpretentiousness; but Schumann is at his happiest when he is suggesting another musical instrument. The Romantic predilection for the horn and its associations with hunting, with the life of the forest and knightly adventure, finds expression in innumerable songs of which 'Waldesgespräch' (Op. 39), 'Der Knabe mit dem Wunderhorn' (Op. 30) and the late 'Der Gärtner' (Op. 107) are only typical examples among many others. The organ-like piano part of 'Stirb', Lieb' und Freud' ' (Op. 35) suggests the cathedral setting of the poem, and the harp-player's broken chords are discreetly suggested in 'Wer nie sein Brot mit Tränen ass' and 'Wer sich der Einsamkeit ergibt' (Op. 98a). In a more humorous vein are the suggestion of the military band in 'Husarenabzug' (Op. 125), the brilliant

EX.2

imitation of a wheezy concertina in the first of the 'Der arme Peter' (Op. 53) songs and the guitar in 'Der Contrabandiste' (Op. 74). Occasionally Schumann writes a whole dance movement where the poem demands it, as in 'Das ist ein Flöten und Geigen' (No. 9 of the *Dichterliebe*) and 'Der Spielmann' (Op. 40), but even here the voice is never overriden.

If we had not his own word for his dislike of Spohr's chromaticism, we might be surprised by the very sparing use of chromatic melody or even strongly chromatic harmony which we find in Schumann's songs. A considerable number suffer from a tendency to the opposite extreme, unrelieved diatonic harmony only too often combined with a square march rhythm, also unrelieved. This can generally be traced to Schumann's desire to write something in either ballad or folk-song style (e.g. 'Sonntags am Rhein' Op. 36) or else to an attempt to recapture the carefree, youthful atmosphere of *Des Knaben Wunderhorn* ('Freisinn' (Op. 25), 'Frühlingsfahrt' (Op. 45) or the 6/8 swinging march of 'Wanderung' (Op. 35) and 'Der Knabe mit dem Wunderhorn'). This rhythmic monotony was already a noticeable trait of some of the longer movements of the piano works (the last of the *Etudes symphoniques* for example) and a tendency to square-cut rhythm remained with Schumann throughout his life. His rare use of chromatic sequences and harmony in the songs is often dictated by the text and he obviously felt very strongly the atmosphere of extreme melancholy, even verging on despair, which such passages produced. In this he was still a child of the eighteenth century. On the other hand his love of quick transitions of mood made him on the whole avoid complete songs in what he felt to be this exaggeratedly melancholy manner. 'Aus den hebräischen Gesängen' and 'Einsamkeit' (Op. 90), with their strongly chromatic, melancholy opening sections, both have comparatively diatonic middle sections in the major key; and only 'Zwielicht' (Op. 39), which is considerably shorter, is allowed to remain in the mood of unrelieved, chromatically coloured gloom throughout. This sparing use of what he plainly felt to be an extreme atmospheric effect makes it all the more effective, and on the very few occasions on which Schumann falls into a strongly chromatic manner in order to express anything but melancholy the result is unpleasantly saccharine and reminiscent of the worse hymn-tunes of John Bacchus Dykes. This resemblance is strongest perhaps in 'Wehmut' (Op. 39) where both tempo and rhythm suggest the hymn-tune; but 'Nur ein lächelnder Blick' (Op. 27), where the chromatic alterations in the melody are combined with a slow 6/8 rhythm and an abysmally vapid poem, may be considered the parent of countless Victorian drawing-room songs which delighted our grandparents but arouse nothing but distaste in us.

The quality which is most typical of Schumann's songs, his most individual contribution to the development of the German *Lied*, is really a noble variety of this sentimentality, the lily which festering in 'Nur ein lächelnder Blick' smells more rank than any weed. The German word for it is 'Innigkeit' and it is virtually untranslatable by any single English word. 'Innigkeit' is a variety of warm, intimate and meditative emotion, essentially self-conscious and therefore dangerously closely allied to sentimentality but saved, at least in its nobler manifestations, by a genuinely childlike simplicity. When this simplicity is also self-conscious, its childlike language becomes something like baby-talk and 'Innigkeit' is then indistinguishable from archness and sentimentality. It is remarkable how many of Schumann songs bear the superscription 'Innig' ('Nur ein lächelnder Blick' among them) and how typical, for good and for evil, these songs are in almost every case—'Widmung', 'Schöne Fremde' (Op. 39), 'Was will die einsame Träne' (Op. 25) for example on the credit side and 'Frage' (Op. 35), 'O ihr Herren' (Op. 37), 'Liebesbotschaft' (Op. 36) and 'Er, der Herrlichste von allen' on the debit, among many others of both complexions. On the whole it is the 'innig' side of Schumann that dates his music most noticeably, just as it was this which endeared him to our grandparents. The circumstances in which the 1840 songs were written—on the eve, or immediately in the wake, of an extremely happy marriage—sometimes give their lyricism a domestic, conjugal quality which may strike the modern listener as a little complacent. Certainly the humble adoration of the chosen male that breathes from the poems of Chamisso chosen by Schumann for his *Frauenliebe und -leben* wakes very little echo in a modern listener. It reflects rather the feelings that the nearly forty-year-old Chamisso was delighted to find in his eighteen-year-old bride. But to a modern taste there is something supremely

unattractive in the poet making his bride speak of herself as a
'nied're Magd' (lowly maiden) who only asks to gaze on her
husband in all humility ('nur in Demut ihn betrachten') and is
blinded by the beauty and amazed at the condescension of so
superior a being. We feel that to be false, possibly quite unjustly;
but if not false, then at any rate the prelude to a 'Frauenleben'
bounded by the three Ks—'Kirche, Kinder, und Küche'. It is
perhaps significant that the second song of the cycle ('Er, der
Herrlichste von allen') is not only the most abject in sentiment
but also the least successful musically, with its square dotted
rhythms and hammered accompanying chords extending un-
interruptedly over four pages. Rückert's two poems in a similar
vein ('Lieder der Braut', Nos. 1 and 2 (Op. 25)) show the girl
torn between her love for her mother and her husband. Here
again a modern reader of the poems might divine a certain
gloating over the situation, comparable with the excessive
interest shown by elderly spinsters in the details of a wedding-
day programme or the exact disposition of bedroom furniture in
the household of a newly-married couple. But Schumann
extracts the last ounce of sentiment from the poems, down to
the sighing 'lass' mich' with which No. 2 closes, without self-
consciousness and therefore without offence.

Nevertheless, it is a very welcome change to move on to the
quite unmatrimonial sentiments of Heine, to find for a change
the male sunk in hopeless adoration and to savour the caddish
and venomous revenge reserved for the last line or couplet of a
poem that seems to start in tremulous humility. Not that
Schumann ever achieves real venomousness; to see how far he
falls short of the possibilities we have only to compare his
setting of 'Anfangs wollt' ich fast verzagen' (Op. 24), a solemn
and resigned chorale, with Liszt's wonderful dramatic miniature.
Nor did Schumann attempt to set 'Vergiftet sind meine Lieder',
which is perhaps Liszt's masterpiece; the venomous quatrain in
the style of Martial could hardly have appealed to the composer
of *Frauenliebe und -leben*. Even so, the lyricism of the
Dichterliebe is a far lighter, more mercurial, more musical
lyricism than anything that Chamisso ever achieved, for all his
French blood. Of the nine songs of the Heine *Liederkreis* (Op.
24) only two come up to the level reached by all sixteen of the
Dichterliebe. The two sets are virtually contemporary, the

Liederkreis songs composed between 1–9 May and the *Dichter-liebe* begun on 24 May and finished on 1 June; so that the difference in quality can only be connected with the choice of poems. 'Ich wandelte unter den Bäumen' and 'Lieb Liebchen' (Nos. 3 and 4 of the *Liederkreis*) would not be out of place in the *Dichterliebe* any more than the four songs which originally belonged to the set but were only published at the very end of Schumann's life or posthumously. These are 'Dein Angesicht' (Op. 127 No. 2), 'Es leuchtet meine Liebe' (Op. 127 No. 3), 'Lehn' deine Wang' ' (Op. 142 No. 2), and 'Mein Wagen rollet langsam' (Op. 142 No. 4, wrongly said by Dr Ernest Walker to be Schumann's last song: it was actually composed on 29–30 May 1840).

What distinguishes all these Heine settings from the remainder of the *Liederkreis*, and from all but a few isolated songs spread over the rest of 1840 and the rest of Schumann's life, is not only their intensity of feeling but their economy of expression. Already in 'Ich wandelte unter den Bäumen' (Op. 24) we meet for the first time that astonishing blend between the simplest folk-song manner and the most subtle and highly organized psychological suggestion, which is the distinguishing note of the *Dichterliebe*. Schumann begins the voice part with a perfectly simple diatonic melody, rhythmically devoid of any subtlety whatever; the accompaniment follows the voice exactly. But after two lines, 'da kam das alte Träumen, und schlich mir in's Herz hinein', the 4/4 rhythm is interrupted by an inimitable triplet phrase which is twice echoed in the piano part (Ex. 3). The second quatrain repeats this pattern and then, with a sudden change—not really a modulation in spite of the dominant minor

EX.3

und   schlich mir in's Herz   hin - ein

ninth—the key shifts from G major to E flat major and the triplet rhythm returns, though at a slower tempo, as the birds offer to tell the poet the magic word they have overheard, the secret of his love. But he will not be robbed of his grief and the folk-song melody returns, only to dissolve into a kind of arioso recitative at the twice repeated 'ich aber niemanden trau'' which dies away in a whisper. The eeriness of the whole poem arises from the suggestion of the 'secret' which the poet refuses to be told, although it would lighten his grief. It is this which gives the neurotic, 'psychoanalytical' atmosphere to so many of Heine's poems, the suggestion of a split personality and a self-torturing, masochistic rapture beneath the conventional roses and nightingales. Here, in 'Ich wandelte unter den Bäumen', Schumann juxtaposes but does not mingle the two personalities. In the *Dichterliebe* his method is far more subtle, for he makes the two 'persons' speak at the same time. Heine plainly enjoyed the torments of unrequited love—'Liebesleid und Weh'—quite as much as the pleasures of mutual passion, which were the only thing that interested Chamisso. (How profoundly disturbed Chamisso was by any less conventionally happy amorous situation is shown by his pathetic poem 'Was soll ich sagen?' (Op. 27), which Schumann seems to have set without really understanding its significance. Chamisso remained Frenchman enough to be mortally embarrassed by an amorous, as opposed to a specifically erotic, relationship between an old man and young girl.)

Schumann, with instinctively right judgement, chose miniature poems of Heine, almost all of them in the extremely simple style of the folk-song. Only one—'Die Rose, die Lilie' (Op. 48)—is as simple in matter as it is in manner and Schumann's setting is as natural and unclouded as the poem. In all the other poems Heine's simplicity of manner is deceptive and what begins apparently as a straightforward lyric assumes before the end the complexion of an enigma or a satirical epigram. 'Wenn ich in deine Augen seh'' (Op. 48), for example, starts like a folk-song and continues conventionally enough until the last couplet. Then, instead of the lover's happiness being full when the girl confesses her love for him, we find the exact opposite, tears: 'Doch wenn du sprichst: ich liebe dich! so muss ich weinen bitterlich.' Schumann starts in a straightforward G major, modulating to the subdominant by means of supertonic minor harmony (which we shall see later to be typical). The first hint of some strange fly in the ointment is the chord of the diminished seventh on the word 'sprichst' but, by a stroke of genius, the actual words 'weinen bitterlich' are in no way underlined harmonically and only the piano postlude gives the hint of something mysterious and uncompleted, by converting the tonic of G major immediately into the dominant seventh of C major so that the tonality of the whole postlude hangs ambiguously between the two keys of C and G.

Sometimes the piano part alone suggests the mystery beneath the poem's surface from the very beginning. The chord of the German sixth with which 'Am leuchtenden Sommermorgen' (Op. 48) opens and the syncopated accents throughout the piano part mentally prepare the listener for the wonderful modulation to the key of G major and the close, once again on the German sixth, which Schumann discovered for the gentle, enigmatic entreaty of the flowers: 'Sei unsrer Schwester nicht böse, du trauriger, blasser Mann.' Finally, in the postlude, the picture of the garden is completed, and over the slowly rippling arpeggios another voice rises, this time the piano's, answering with wordless consolation the unspoken complaint of the poet. I personally should place this one song among the very greatest miniatures in the whole of music, a faultless dramatic lyric (with the lyricism unbroken on the surface and the drama implicit), the *locus classicus* of 'Innigkeit' at its very best.

The two dream poems of the *Dichterliebe*—in which we should naturally expect the strange and pathological element to predominate—are in marked contrast to each other. 'Ich hab' im Traum geweinet' owes a large part of its effectiveness to the long pauses, the unaccompanied phrases of the voice, and the sinister dotted quaver figure in the piano part which punctuates the lines of the poem (cf. Duparc's 'Le Manoir de Rosamonde'). Once again we find extreme agony of mind expressed in the chromatic harmony of the accompaniment to the last couplet, where the voice part ends on a chord of the dominant and out of key (dominant of A flat in the key of E flat minor). The dream is precise, bitterly clear in every detail. 'Allnächtlich im Traume' with its short, gasping phrases and sudden changes of rhythm is the dream which remains clear only as a general impression, all

the details being blurred. Schumann's setting has the breathless-
ness of Heine's poem, the sudden, childish spurts of confidence,
ending with the pathetic, frustrated confession, 'und's Wort
hab' ich vergessen': the whole point of the dream has eluded
him. Singers are apt to take the song too slowly and to make it
an expression of bliss, when the sense of both words and music
demand an atmosphere of restlessness and anxiety. Schumann
never modulates properly to the key of the dominant, either in
this song or in 'Ich grolle nicht'. If he reaches the dominant he
moves away from it at once, after a single beat, preferring
subdominant harmony or, in 'Ich grolle nicht', a maximum of
variety without actually leaving the tonic key at all. This is the
only one of the *Dichterliebe* cycle with a repeated-chord
accompaniment, which is justified by the shortness of the song
and the urgent, persistence of the words, the quick mounting to
a thunderous climax (where, by the way, the alternative, higher
voice part was added by the composer only as an afterthought
when the song was in proof).

    Hardly less of a dream is the last song but one of the set, 'Aus
alten Märchen', one of the most successful of Schumann's 6/8
march songs, with the persistent rhythm mitigated by considerably greater harmonic variety and changes in the weight of the
accompaniment than in other songs of the same kind. The
augmented version of the melody in the last verse, marked 'mit
innigster Empfindung,' concentrates the whole weight of the
song in the close and makes the sudden evaporation of the
whole vision in a series of diminished sevenths both more
unexpected and more effective. 'Die alten, bösen Lieder,' which
ends the set, opens with a dramatic flourish that is very rare in
Schumann's songs, and the piano part—with its marked
resemblance to No. 4 of Chopin's *Etudes*, Op. 10, which had
appeared in 1831—bears the main burden of the musical
interest until the, by now familiar, concentration of energy and
intensification of emotion which herald the final quatrain. The
weighty, mounting octaves in the bass of the piano part recall a
similar passage in the first movement of the *Eroica* (probably
intentionally), but when the voice has finished the octave
portamento rise on the C sharp, the whole mood of the music
changes in a moment from brutal violence to tremulous self-
pity. This is the reverse of the usual Heine process; the sting is

not in the tail of the poem, for once. Instead of a savage or enigmatic close Schumann has dramatic justification for one of his most cherished effects—the sun suddenly bursting through the clouds, a wave of tenderness bursting in upon harsh sarcasm. He achieves this by an enharmonic modulation pivoting on the C sharp, which is suddenly treated as the leading note in the key of D instead of the dominant of F sharp. This lasts only two bars and the voice part dies away over a dominant harmony in the original key of C sharp minor. Then the consoling voice of the piano rises—exactly as in the postlude to 'Am leuchtenden Sommermorgen'—only here the postlude is followed by a further instrumental passage in the nature of an improvisation which rounds off the whole cycle. A descending quaver passage in the right hand is developed sequentially and harmonized with rich chromatic harmony until it finally dies away in a thrice-repeated sigh.

No other poet provided Schumann with the concentrated intensity of feeling and the verbal terseness and economy that were the two things he most needed from a poem. Or if an occasional lyric (Mörike's 'Das verlassene Mägdlein' (Op. 64) or Ulrich's 'Die Fensterscheibe' (Op. 107), for example) has these qualities, we still miss the mysterious, charged atmosphere of the Heine settings. Nos. 11 and 12 of Op. 35, 'Wer machte dich so krank?' and 'Alte Laute', have something of the same quality; but Justinus Kerner, the poet, was a doctor and an occultist and the mystery behind the poems is objectively imagined rather than subjectively experienced. This accounts for the tame ending of 'Alte Laute' with the introduction of the conventional figure of the angel and the correspondingly conventional cadence in the music. Far more typical of Schumann's general lyrical manner is No. 10 of the same Op. 35, 'Stille Tränen'. Kerner's poem is one of the innumerable lyrics inspired by the secret grief of the poet and the uncomprehending attitude of the world, a Romantic commonplace which he treats no better and no worse than a hundred other minor poets have done. It lacks entirely the drive and bite that Heine's sense of humour and power of self-criticism give to even his most effusive poems; and Schumann's music, for all the beauty of the melody and the richness and subtlety of the modulations, is too discursive and rambling. His repetition of the last couplet, preceded and followed by an interlude and postlude of comparative lengthiness, represents a third of the whole song, which is thus artificially enlarged. There is no case of Schumann's treating a Heine poem in this way, though it becomes almost a mannerism in his settings of other poets. Even the beautiful No. 1 of Op. 39 (the Eichendorff *Liederkreis*), 'In der Fremde', repeats the last line simply for the sake of repeating the effective cadence (a G natural in the key of F sharp minor). The other song of the same name in the same set ('Ich hör' die Bächlein rauschen') repeats the last line of the poem three times without even the excuse of a harmonic or melodic *trouvaille*, and in a really poor song such as 'Liebesbotschaft' (Op. 36 No. 6) the general rambling and sectional character of the whole setting is enhanced by a spate of arch repetitions at the end.

Apart from five poems in the *Myrthen* (Op. 25) of 1840, Schumann did not attempt to set any but incidental poems by Goethe until the nine songs from *Wilhelm Meister* (Op. 98a). Goethe's balance and serenity, philosophic depth, and dislike of romantic exaggeration all made him a quite unsympathetic figure to the young Schumann; and later Schumann's attempt to widen his horizon and to achieve a musical language capable of expressing Goethe's thought led him, at least as a song-writer, to go against all his natural instincts—or rather, since instinct generally gets the better of the artist, to give free rein to the most unfortunate of his natural tendencies, rambling and divagation. The lyrics from the *West-östliche Divan* in *Myrthen* are charming and effective even if only one of them, 'Lied der Suleika', really catches the spirit of Goethe; and even there the otiose repetitions are a serious blemish in the setting of such a formal master as Goethe. The ballad of 'Die wandelnde Glocke' (Op. 79) is one of the better of Schumann's songs in this style. It attracted him from a literary and theoretical point of view but was in no way suited to his essentially lyrical and subjective temperament. The *Wilhelm Meister* songs, on the other hand, are among Schumann's most conspicuous failures as a song-writer. Painfully oppressed by the philosophic significance of Mignon and the old harp-player, he rambles on in a portentous pseudo-symphonic style, with frequent modulations and un-natural vocal phrases, losing the thread of the poem and of his own musical design, and sometimes, as in No. 6, 'Wer sich der

Einsamkeit ergibt', visibly at a loss how to continue (Ex. 4). Only Philine's 'Singet nicht in Trauertönen' is at least half successful, though here again the repetitions of the last line, and then again of the last phrase, are unforgivable.

EX.4

In addition to settings in ballad style scattered all over his sets of songs, Schumann wrote four sets with the definite title of *Romanzen und Balladen*, Opp. 45, 49, and 53 in 1840, and Op. 64 in 1847. These contain sixteen songs in all, with texts by Heine, Eichendorff, Mörike, Seidl, Lorenz and Fröhlich. In the same category are Heine's 'Belsatzar' (Op. 57)—though actually one of the first of the February 1840 songs—Schiller's 'Der Handschuh' (Op. 87), and the melodramas or ballads for declamation—Hebbel's 'Schön Hedwig' (Op. 106), and 'Ballade vom Haideknaben' (Op. 122 No. 1), and Shelley's 'Die Flüchtlinge' ('The Fugitives') (Op. 122 No. 2).

Of the three poems by Chamisso (Op. 31) two are definite ballads, 'Die Löwenbraut' and 'Die rote Hanne', and the third,

'Die Kartenlegerin', is a romantic genre picture in much the same style. Two of the Heine settings are miniature dramas consisting of three poems each, 'Der arme Peter', (Op. 53 No. 3) and 'Tragödie' (Op. 64 No. 3). They are very unequal in quality and only one song from each triptych is comparable with Schumann's best settings of Heine. The concertina which accompanies Grete's dance with Hans, poor Peter's rival ('Der arme Peter' No. 1) turns the whole song into a slow country waltz or *Ländler*. Peter's misery finds no expression in the music; he is merely part of the picture, in the true ballad style. The second song in 'Tragödie', on the other hand, is lyrical in character and only successful in comparison with the fatuity of the first, and the sentimental banality of the third parts. 'Belsatzar' starts rather unpromisingly like a piano solo—in fact the semiquaver figure in the piano part which dominates half the song is closely related to a passage in No. 5 of the *Fantasiestücke*, Op. 12 ('In der Nacht'). The great length of the poem, eleven verses of four lines each, made it extremely difficult to avoid monotony; but Schumann succeeded here as he seldom did in later ballad settings. This is chiefly owing to the way in which the piano part becomes increasingly simple and tends more and more to make way for the voice as the dramatic interest of the poem increases. The semiquavers, which are almost continuous in the first half of the poem, give way first to repeated quaver chords, with the accent off the beat; and then to isolated chords merely supporting the voice in the last two verses, where the tempo decreases and the chief point lies in the dramatic recitation of the text ('In langsameren Tempo, leise und deutlich zu recitiren'). The glaring fault of both 'Die Löwenbraut' and 'Blondels Lied' (Op. 53), rhythmic monotony, was thus avoided in 'Belsatzar' as in the lyrical ballad of 'Die beiden Grenadiere' (Op. 49). In 'Die Löwenbraut' there are forty continuous bars of slow 3/2 time, unrelieved by any variation in the piano pattern, which follows the voice exactly. Similarly in the 124 bars of 'Blondels Lied'—'nicht schnell' 4/4—there are only twenty bars in which the piano part does not follow the voice exactly; often it is in octave unison or in the simplest diatonic harmony devoid of harmonic, melodic, or rhythmic interest. Schiller's 'Der Handschuh' is an exciting story told with the obvious and rather selfconscious dramatic effect we should expect. Schumann amuses himself by some equally naïve dramatic effects (major and minor ninth intervals in the voice when the lion roars and lashes his tail) and there is plenty of variation of key and rhythm, so much so in fact that musically the song does not really cohere at all. It is in fact a hybrid between a real ballad such as 'Belsatzar' and the declaimed melodramas, in which the actor plays the main role and the music is merely 'background'. Unfortunately this is a role that music obstinately refuses to play and the melodramas are complete failures, though the 'Ballade vom Haideknaben' is an exciting story and Schumann's illustrations are often apt enough.

At the opposite extreme to the ballads stand Schumann's salon pieces. They are not many, for the simple reason that Schumann was never by nature a frequenter of salons. Domestic, conjugal felicity, which speaks rather too smugly for modern taste from some of the songs, satisfied both his emotional nature and his demands for the society of his fellow creatures. Nevertheless, every middle-class drawing-room in the Germany of the 1840s and fifties had its cultural pretensions and was, indeed, a miniature salon, if only for the family circle. Albums, pressed flowers, needlework, water-colours, piano duets and, of course, singing all contributed to this specifically domestic form of culture; and though it was only in exceptional salons that these rose much above the status of genteel crafts or achieved that of arts, the demand on all artists to produce at least occasional pieces for home use was continuous and, as their works show, fairly continuously met. The works of even such a comparative misanthrope as Grillparzer contain innumerable *pièces d'occasion* 'for Miss X's album', 'on a water-colour by Miss Z', and the like, while a socially-inclined artist like Liszt must have had the greatest difficulty in refusing demands which, if satisfied, would have taken up all the time he had for composition. As it is, a large proportion (25 out of 57 in Kahnt's three volumes) of Liszt's songs are settings of elegant trifles written by friends, of both sexes, belonging to the beau monde. Schumann was probably less plagued than many artists because of his inclination to solitary brooding, his frequent 'absence' in social gatherings; but it would have been unnatural if there were not some salon pieces among his songs. Indeed, his gifts as a miniaturist and his lyricism, though rather intense and highly charged for the drawing-room, fitted him for this minor genre, and there are at least two models of the kind among the 1840 songs, Rückert's 'Jasminenstrauch' (Op. 27) and Catherine Fanshawe's 'Rätsel' (Op. 25) (the poem itself consisting of a parlour game and wrongly attributed to Byron). Far more typical of the general level of such works is the song 'Liebste, was kann denn uns scheiden', No. 6 of the poems from Rückert's *Liebesfrühling* (Op. 37) which Schumann and his wife collaborated in setting. The four verses of the poem, each starting with the question, 'Beloved, what can part us?', continued with verbal assonances on 'meiden' and 'scheiden', 'mein' and 'dein', and the obvious variations of an arch party-game. (It is worth comparing the song with Liszt's setting of Charlotte von Hagen's 'Dichter, was Liebe sei, mir nicht verhehle', a more personal and dangerous variant of the same game.) The musical interest is virtually non-existent. Both 'Röselein, Röselein' (Op. 89) and 'Schneeglöckchen' (Op. 96) are all more highly developed, more sophisticated salon pieces; and even 'Stille Liebe' and 'Aufträge' are not really much more. The arch 'question' in the piano part with which 'Stille Liebe' opens (Ex. 5), with its superscription 'innig', is comparable with the teasing triplets at the opening of Liszt's 'Dichter, was Liebe sei', though the emphasis is all on emotion in Schumann and on grace in Liszt. In the same way the rippling demisemiquavers that represent the wave in 'Aufträge' are a great artist's version of what is in origin no more than a parlour trick.

EX.5

Among these amusements and accomplishments of the nineteenth century, drawing-room duet-singing had a place which was half artistic and half (shall we say?) social. Music, as Plato knew, is an excellent dissolver of inhibitions and barriers of all kinds and if Edward could persuade Bertha to learn a duet with him he could reckon on the power of both music and text (generally discreetly amorous), the appeal of his own charming tenor, and his exquisite tact and consideration in turning pages, modifying his tone so as not to drown her pretty soprano, the necessity of looking one's partner straight in the eyes so as to ensure a perfect entry of both voices together and a hundred and one delightful details which gave courtship all the excitement of an honourable adventure and a good many of the thrills of the chase. Nineteenth-century duets, then, are not to be judged on purely musical grounds. Like many Russian symphonies, they were strictly *Gebrauchsmusik* and, as such, the capabilities of the performers must not be rated too high by the composer. Of Schumann's thirty-four duets only a handful are of any interest musically. The majority are comparable in interest of design, as in original intention, with the small form of sofa called a *causeuse*, large enough for two people sitting very close together and making no violent movements. Typical of this sort of duet are 'Liebesgarten' (Op. 34) (which looks like a banal sketch for 'Im Walde' (Op. 39), though it was actually written later) and the duet numbers from Rückert's *Liebesfrühling*, including the duet version of 'Liebste, was kann denn uns scheiden?'. 'Familiengemälde' (Op. 34), with its sentimental comparison of the young couple and the old Darby and Joan, is a typical conversation-piece in Schumann's most bourgeois manner. The four duets (Op. 78) contain some much better music. The 'Tanzlied' is a charming drawing-room game, needing considerably more musical ability than anything of the kind that Schumann had written previously; but I suspect that the boldness of the text ('morgen, o Trauter, dein ganz') confined performances to professional singers in any case. This applies also to the two 'vergebliche Ständchen', 'Liebhabers Ständchen', and 'Unterm Fenster' (Op. 34). The setting of Goethe's 'Ich denke dein' (Op. 78) keeps the voices singing mostly in thirds and sixths like all the salon pieces, but it has a pretty romantic sentimentality and must have been a powerful weapon in Edward's amorous arsenal. 'Wiegenlied' (Op. 78) (Cradle-song for a sick child) has not the mawkish flavour we might expect and though the relation between the two voices is still very primitive and uninteresting, the contrast between triplets and plain four-in-a-bar quavers, the sudden modulations from E minor to C major and from C to E flat, and the exploitation of the passing note (A sharp in the E minor triad) makes the song unique among the simpler salon duets of Schumann. The duets in the *Liederalbum für die Jugend* and the settings of Elisabeth Kulmann (*Mädchenlieder* (Op. 103)) are of no interest musically.

There remain the two sets of Spanish pieces, *Spanisches Liederspiel*, Op. 74 (five duet numbers), and the two duets in the *Spanische Liebeslieder*, Op. 138. Spanish pastiche attracted Schumann rather in the same way that Hungarian gipsy pastiche attracted Brahms. In neither case was there any attempt at understanding a different musical tradition, and the resemblance of the result to the original is only occasional and superficial. Nothing further was aimed at, of course, but even so Schumann's Op. 74 does contain some very extraordinary things. Of the duets 'Erste Begegnung' and 'Liebesgram' have vaguely exotic rhythms, strongly dotted, that is to say, with occasional accents off the beat and some 'passionate' triplets, exactly like Brahms's so-called Hungarian pieces. 'Intermezzo' (Op. 74) is a very German serenade, more like a cradle-song imitated from German folk-song, such as Brahms affected later. 'In der Nacht' again suggests Brahms's future style, in its most un-Spanish combination of Bach-like chromaticism (Ex. 6) and sweet, warm German sentiment. Finally 'Botschaft' opens with a direct quotation from the middle section of Chopin's C sharp minor Polonaise, Op. 26, and makes considerable demands on the

EX.6

singers in the two really independent vocal lines which only
come together for florid ornamentation with wide intervals (Ex.
7). This same tendency to wide intervals is found in one of the
solo numbers of the *Spanisches Liederspiel*, 'Melancholie', a
strange and violent little song, again with Bach-like character-

EX.7

istics (Ex. 8) and containing dramatic leaps of ninths, tenths,
and even in one place a fourteenth, which are most unusual in
Schumann's vocal writing. 'Der Contrabandiste', with its guitar
accompaniment, coloratura and triplets imitating a horse's

EX.8

gallop, is really a parody and quite an amusing one. The guitar is suggested again in the 'Flutenreicher Ebro' from the *Spanische Liebeslieder*, Op. 138, whose title and pianoforte duet accompaniment obviously gave Brahms the idea for his own *Liebeslieder*. None of these solos or duets bears the remotest resemblance to anything genuinely Spanish, and the general musical level of interest is below that of the earlier *Liederspiel*. Thus 'Bedeckt mich mit Blumen' is rather like the earlier 'Liebesgram', only less successful, and 'Blaue Augen hat das Mädchen' is a much less interesting essay in the German folksong style than 'Intermezzo'. Of the solo songs 'O wie lieblich ist das Mädchen' and 'Hoch, hoch sind die Berge' are essays in much the same style and their chief distinction lies in the fact that, once again, Brahms (I suspect) had a passage from 'Hoch, hoch sind die Berge' (Ex. 9) in mind when he wrote 'Vergebliches Ständchen'.

EX.9

In between the ballads and the salon pieces, and partaking of the nature of both, come first the *Liederalbum für die Jugend*, Op. 79, and the *Sieben Lieder*, Op. 104, settings by the child 'poet', Elisabeth Kulmann. Like the *Kinderscenen* for piano, these songs are written for children as they exist in the sentimental imagination of an adult rather than for children as they are. It is possible that in the nineteenth century children could be persuaded to adopt, at least in public and before their parents, some of the qualities attributed to them in the songs—extreme *naïveté*, love of nature, a sense of pity and at least nascent Sabbatarianism. But that any children, in any country, at any date, should really have been anything but complicated, indifferent to nature, pitiless and impious by nature, I cannot believe. Schumann, then, either writes for the 'Kunstkind', the product of intensive nursery training, or not for children at all. 'Schmetterling', 'Frühlingsgruss', 'Sonntag', 'Hinaus ins Freie', 'Weihnachtslied', and 'Kinderwacht' are all carefully simple musically, unctuous in tone and quite devoid of any interest. But spread among these thirty-odd songs there are some exquisite miniatures. The first of the two 'Zigeunerliedchen' (Ex. 10), for example, and 'Der Sandmann' have a lightness and aptness of touch, a sureness of psychological instinct shown in minute detail, such as only Hugo Wolf has equalled. They are neither folk-songs nor children's songs but spring from a spontaneity of feeling and directness of expression which Schumann at times possessed in common with 'primitive' people and children. No child ever thought or felt in terms of Hoffmann von Fallersleben's 'So sei gegrüsst vieltausendmal, holder, holder Frühling'

EX.10

Un – ter die Sol – da – ten ist ein Zi – geu – ner – bub' ge – gan – gen

but any child might prick up its ear at a story which began 'Unter die Soldaten ist ein Zigeunerbub' gegangen'. 'Des Sennen Abschied' from Schiller's *William Tell* might be taken as a companion piece to Lenau's 'Die Sennin' (Op. 90 No. 4) as both depict the descent in the autumn from the high mountain pastures of the Alps. Schiller's poem is the shepherd's own song and Schumann's accompaniment suggests a primitive pipe-tune which gradually dies away in the distance. Lenau's is a German equivalent of Wordsworth's 'Highland Reaper', a deeply subjective lyric full of Ruskin's pathetic fallacy (the mountains miss the girl and remember her songs). The harmonic scheme is correspondingly more rich and complicated, and whereas 'Des Sennen Abschied' only modulates occasionally to the dominant or the relative minor of its tonic C major, the voice part of 'Die Sennin' ends on a chord of the dominant, while the piano part dies slowly away on a mediant major chord (D sharp major in B major).

The three Goethe settings which have somehow found their

way into the *Liederalbum für die Jugend* are quite out of keeping with the rest of the contents. Lynceus's song from *Faust*, Part 2, is a monotonous piece in the four-square dotted rhythm beloved by Schumann, the setting of 'Kennst du das Land' obviously belongs to the other *Wilhelm Meister* songs in Op. 98a, where it also appears. It cannot compare with the setting, however faulty, by Liszt or the magnificent one by Hugo Wolf, but it is very much better than the rest of the *Wilhelm Meister* songs and interesting for Schumann's very rare insistence on the chord of the dominant minor ninth ('Was hat man dir, du armes Kind, getan?').

Schumann's admiration for the poems of Elisabeth Kulmann, which issued in his Op. 104 (1851) is explicable partly as the result of failing powers of judgement, perhaps, but partly also as one of those quite frequent lapses in taste which may occur to any artist of a deeply emotional character liable to be influenced by personal circumstances. Elisabeth Kulmann was born at St Petersburg in 1808 and lost her father and six of her seven brothers while she was a young child. She herself died of consumption in 1825, after living in poverty with her mother. Her poems show her to have been a sensitive and idealistic young woman, whose imagination was probably stimulated by the disease from which she suffered. She had a poetic facility which enabled her to imitate the lighter, occasional lyrics of the Romantic school with a certain elegance, and a spiritual quality which showed itself in the simplicity and resignation with which she faced death. Schumann was greatly struck by the unusual pathos of her story and, in an introduction which he printed to his settings of seven of her poems, he went so far as to prophesy her universal recognition as a great German poet and to speak of her poems as 'wisdom's highest teachings expressed with the poetic perfection worthy of a master'. It must be remembered that his literary taste was always very personal rather than aesthetically correct, and that to the end of his life he rated Jean Paul as the equal of Goethe and Shakespeare. The settings of Elisabeth Kulmann's songs witness to the soundness of his musical instinct, in any case, for the unpretentiousness of the music exactly matches the poems, with their maidenly lisping of swallows, greenfinches, dead flowers, clouds, and the moon. They are of no intrinsic musical interest whatever.

It is an interesting fact, which says much for Schumann's understanding of the nature of vocal music, that the rhythmic complications, anomalies and puns which are such a marked feature of his pianoforte style seldom, if ever, appear in the voice parts of his songs; and hardly ever oppressively even in the piano parts. In fact, as we have already seen, a considerable proportion of the songs suffer from the opposite fault, monotony and jejuneness of rhythm. Schumann always tended to confine his rhythmic experiments to detail, often to inner parts, while the movement as a whole—whether it be orchestral, vocal, or chamber music—progresses in uniform, scarcely disguised march time. His most completely successful pieces are generally miniatures and, in the songs at least, the subtlety and effectiveness of his rhythm is generally in inverse proportion to its complexity. Compare, for instance, 'Die Fensterscheibe' with 'Abendlied', both from Op. 107. Schumann very seldom changes the time signature in the course of a song. When he does—as in 'Allnächtlich im Traume' or 'Jemand' (Op. 25)—it is for a quite definite dramatic purpose. In some of the most successful songs the piano continues the same rhythmic pattern from beginning to end ('Der Nussbaum' (Op. 25), 'Mondnacht' (Op. 39), 'Frühlingsnacht' (Op. 39), and most of the *Dichterliebe* songs) and the voice has its own, perfectly distinct melody which is either completed or commented on by the piano ('Nussbaum') or leads an entirely separate existence ('Das ist ein Flöten und Geigen'). A note-by-note accompaniment, with the top note of the piano part permanently in unison with the voice, is sometimes skilfully masked by ornamentation, as in 'Und wüssten's die Blumen, die kleinen', (Op. 48) or broken up into brilliant fragments, as in 'Frühlingsnacht'. It is generally only left quite undisguised in the march songs in 'folk' or student style ('Frühlingsfahrt' (Op. 45), 'Wanderlied' (Op. 35) and the rest) and even then not for the whole song in the better examples. Hymns like 'Zum Schluss' (Op. 25), 'Talismane' (Op. 25), or 'Nachtlied' (Op. 96) are of course derivatives of these march songs; but even in these Schumann generally breaks the plodding 4/4 by some device, the flowing quavers in the middle section of 'Talismane' and the festoons of crotchet triplets around the words 'Warte nur!' in the Goethe poem ('Nachtlied'). Occasionally Schumann overworks an idea which is good in itself but loses its initial point by repetition. The piano part of 'Muttertraum' (Op. 40), with its wandering semiquavers and cross-bar suspensions, sounds like a two-part invention to which the voice part has been added afterwards and, though it makes an effective opening, it does not seem to correspond to two whole quatrains of Andersen's poem. In much the same way 'O Freund, mein Schirm, mein Schutz' (No. 6 of Rückert's *Minnespiel* Op. 101) seems rather an essay in the style of Bach—perhaps suggested by Rückert's poem which addresses the earthly much as Bach's cantata arias address the heavenly lover—than a genuinely original inspiration. The exclusive use of passing-note harmony and the same angular rocking rhythm, as of a broken cradle, for nearly fifty bars without a single break is an extraordinary instance of lack of self-criticism in such a self-conscious artist as Schumann. Compare this with a song of much the same length (though admittedly much quicker tempo) like 'Ein Jüngling liebt ein Mädchen' (Op. 48) where Schumann starts with the idea of a strong accent off the beat and indeed carries it right the way through the song but is careful not to emphasize the syncopated rhythm except in the piano solo passages, prelude, interlude, and postlude.

This scrupulousness in his treatment of the poems which he set places Schumann, as we have seen, half way between the unscrupulously musical Schubert, and the almost painfully literary Wolf. His two misquotations ('Blätter' for 'Äste' in Mosen's 'Der Nussbaum' and 'lieblichen' for 'guten' in Heine's 'Das ist ein Flöten und Geigen') are little more than misprints and the only serious criticism that his poets could make of Schumann's settings of their poems is that he tends, for purely musical effect, to repeat their last lines or last phrases. The examples of this are innumerable and I have referred to some already. Even Heine is occasionally mishandled in this way ('Berg' und Burgen' (Op. 24), 'Ich wandelte unter den Bäumen') though generally the construction of his poems makes it virtually impossible without giving the impression of repeating the end of a witty story to be sure that the audience has not missed the point. What led Schumann to offend against literary canons in this way was his desire to impart a dream-like, echo effect to a song, a heightening of emotion as the music dies away in the distance. Sometimes this effect is achieved by a simple repetition of the last phrase, differently harmonized but with no piano postlude, as in 'Im Walde'. Sometimes—and this is the most frequent method—the last line or phrase is repeated and a postlude added as well, as in 'Stille Tränen', 'Nur ein lächelnder Blick', 'Liebesbotschaft' and many others. But the same effect is achieved by purely musical means and without deforming the poem in all Schumann's best songs simply by the piano postlude taken in conjunction with the last line of the poem, as in 'Widmung', 'Der Nussbaum', 'Mondnacht', 'Frühlingsnacht', 'Am leuchtenden Sommermorgen', 'Jasminenstrauch', 'Auf das Trinkglas eines verstorbenen Freundes' (Op. 35), 'Die beiden Grenadiere', 'Die Fensterscheibe'—to choose examples from every date and category. In these songs the musical and literary emphases coincide, instead of Schumann coming to the end of his poem before the music has reached its natural climax, as in 'Stille Tränen'—the most flagrant example of all—where twenty-one bars out of a total seventy are tacked on after the poem has finished, in the form of interlude, repetition of the last couplet, and long postlude.

As we have seen, Schumann was conservative in his use of chromaticism, but this does not mean that the songs are in any way poor harmonically. A few, as we have seen, are too rich and a considerable number suffer from a deliberate simplicity employed, often unsuccessfully but always deliberately, for a definite purpose—the creation of an imaginary 'folk', child, or student atmosphere. None of Schumann's masterpieces come into either of these classes, the over-rich or the artificially slimmed. In fact, the beauty of his most successful songs depends very largely on the perfect accord between poetical matter and harmonic manner. In setting a simple poem such as Mosen's 'Der Nussbaum', Schumann employs a very simple harmonic scheme which does not go beyond the tonalities of the mediant (minor), the subdominant, dominant and supertonic minor. Within these keys there are no further chromatic alterations and the ambiguous chords of the dominant seventh or chromatic alterations of the sixth are instinctively avoided. In an intensely emotional lyric such as Eichendorff's 'Mondnacht', on the other hand, Schumann deploys much more harmonic subtlety. The actual form of the song is simple almost to monotony: the voice part consists of the same eight-bar phrase repeated four times and followed by two four-bar phrases which lead to the fifth and last repetition of the original phrase slightly varied to make a full close. In 'Der Nussbaum', as in many other songs, Schumann tended to avoid the dominant, to shy off it by a harmonic side-slip (the only occasion where the dominant seventh is used) and to prefer the minor of the mediant or supertonic. In 'Mondnacht' the piano prelude modulates to the dominant (B major) within two bars and seems to insist on it until the voice enters and we find ourselves almost immediately in F sharp major, that is to say the key of the supertonic from the point of view of E major (the tonic key) or the dominant from the point of view of B major. This playing with ambiguities of chords of the dominant reaches its climax at the beginning of the last quatrain where, on the word 'spannte' ('spread') an inversion of the dominant seventh in the tonic (E major) suddenly becomes a dominant major ninth in the key of A, as the spreading of the soul's wings is suggested by the wide interval and spreading semiquaver phrase in the piano part. Schumann even ends the voice part on a chord of the dominant seventh, out of the key (A major instead of E) and there are many instances of this, at that time revolutionary, process throughout the songs. In the *Dichterliebe* alone, 'Im wunder-schönen Monat Mai', 'Ich hab' im Traum geweinet', 'Das ist ein Flöten und Geigen' and, 'Am leuchtenden Sommermorgen' end on chords of the dominant seventh out of the tonic; and 'Aus meinen Tränen spriessen', 'Ich will meine Seele tauchen', 'Im Rhein, im heiligen Strome', 'Allnächtlich im Traume', 'Aus alten Märchen', and 'Die alten, bösen Lieder' all close either in the key of the dominant or else on a dominant seventh in the tonic. This applies only to the voice parts, of course, and in most cases Schumann makes the piano postlude come to a more or less conventional cadence in the tonic. Not always, however. In 'Die Nonne' (Op. 49), for example, both voice and piano part close on a dramatic dominant seventh and in 'Die Sennin' the voice ends on a dominant seventh of the tonic and the piano part unmistakably in the key of the mediant.

Occasionally Schumann opens a song with a chord of the dominant or diminished seventh ('An meinem Herzen, an meiner Brust' from *Frauenliebe und -leben*; 'Flügel! Flügel! um zu fliegen' (Op. 37), 'Was will die einsame Träne?', 'Die Tochter

Jephthas' (Op. 95) and even more frequently whole passages in the course of a song are built up on sequences of dominant sevenths. Among the most obvious examples is the quiet middle section of 'Widmung' and the magnificent chain of dominant sevenths which leads up to the climax in 'Aus alten Märchen' ('und bunte Quellen brechen' to the twice repeated 'Ach!'). We have already seen a more sophisticated version of the same thing in 'Mondnacht' and there are two other noticeable examples in the Eichendorff *Liederkreis*: 'Schöne Fremde' and 'Frühlingsnacht'. All these nocturnes or night pieces are in extreme sharp keys, as though Schumann felt that this blend of richness and ambiguity were his own way of expressing the Romantic adoration of the night—Novalis's 'herrlicher Fremdling mit den sinnvollen Augen, dem schwebenden Gange und den zartgeschlossenen, tonreichen Lippen'. More rarely a comparatively diatonic song will close with a postlude which consists almost entirely of a chain of dominant sevenths, for example 'Schöne Wiege meiner Leiden' (Op. 24) (Ex. 11), and 'Rose, Meer und Sonne' (Op. 37).

EX.11

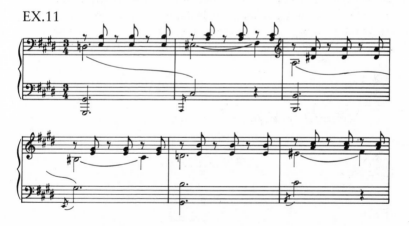

There was nothing unusual in Schumann's use of the dominant seventh; that is to say, there was nothing unusual in the fact that he used it. A far more unmistakable harmonic finger-print of Schumann's is his predilection for—almost his obsession with—supertonic minor harmony (especially, it seems, in the key of A major, where passages in B minor almost always

appear sooner or later). The opening phrases of 'Was will die
einsame Träne' (Ex. 12) and 'Lied der Suleika' (Ex. 1) are typical
instances, but examples are spread thickly throughout songs of
every period. It is dramatically crucial in some of the *Dichterliebe*
songs, at 'da ist in meinem Herzen die Liebe aufgegangen' in 'Im

EX.12

wunderschönen Monat Mai' and at 'ich liebe dich' in 'Wenn ich
in deine Augen seh' '. Again in 'Was soll ich sagen?' we find the
same harmony at the dramatic climax, 'du heissest mich reden'.
Sometimes it is reached through the subdominant, as in 'Stille
Tränen' (C major), where the D minor ('ob allen Landen') is
reached through the dominant seventh of F major, the subdominant of the tonic key; or by means of the dominant, as in

EX.13

'Aufträge'. Sometimes it is a mere passing effect, almost always
emotional, as in 'Heimliches Verschwinden' (Op. 89) or 'Meine
Rose' (Op. 90) (Ex. 13); or it is used as an ornament to give
dramatic emphasis to a phrase, as in the last bars of the voice
part of 'Ihre Stimme' (Op. 96) (Ex. 14).

EX.14

Another favourite and very effective device of Schumann's, which he uses with great care and never wastes, is enharmonic modulation. The roots of almost all his harmonic experiments can be found in Schubert, including the technique of jumping from dominant to dominant which we have just seen, and the turning to the key of the mediant where more conventional composers would have modulated to the dominant ('Zwischen-dominanten' and 'Mediantenrückung', as the Germans call these devices). The pivoting on a single note we have already met in the last bars of 'Die alten bösen Lieder', where Schumann used it to express the sudden change in the poet's mood. In 'Die Lotosblume' (Op. 25), 'Stille Tränen', and 'Alte Laute' there is no dramatic literary reason and Schumann is guided by purely musical instinct. The effect is certainly breathtaking when the A flat major breaks in on the C major at 'Der Mond, der ist ihr Buhle' in 'Die Lotosblume' and at 'So lang du ohne Sorgen' in 'Stille Tränen': and hardly less when the simple, sequential phrases of 'Wer machte dich so krank?' are suddenly transferred from the dominant (E flat) to G flat major, i.e. the key whose tonic is the third note in the minor of the dominant (Ex. 15). This is a sophisticated form of the tendency to mediant harmony which we have already noticed (in 'Der Nussbaum', for example). Although there is no immediate dramatic reason for these enharmonic changes, Schumann nearly always uses them in songs whose general sense and atmosphere have something eerie or uncanny; and this is obviously in accordance with his general feelings, already noticed, of the extreme power and effectiveness of chromaticism in any shape or form. We have

EX.15

und   kei - ne Ster - nen - nacht   Kein Schatt - en un - ter   Bäu - men

already noticed the enharmonic modulations in two songs from
Op. 35, both poems by Justinus Kerner, the doctor and
occultist. Of his poems, No. 6 of Op. 35, 'Auf das Trinkglas
eines verstorbenen Freundes', is an unusual combination of
Schumann's extremely rigid, four-square, dotted rhythm—a
cross between the march and a prayer like 'Zum Schluss'—with
harmonic experimentation which alone suggests to the listener
the mystery of the poem. The piano follows the voice part note
by note for three quatrains, without ornament of any kind apart
from the harmonic colour. At the end of the first quatrain the
music has reached what is presumably the dominant of B flat
major (the song being in the key of E flat). This unison F,
Schumann now chooses to regard as the mediant of D flat major
and a four-bar phrase, repeating exactly the rhythm of the first
two couplets, now follows in that key. This in its turn ends on a
chord of D flat major with the mediant uppermost and is
followed by a four-bar phrase identical with what has gone
before but in the key of F minor (the supertonic minor of the
original E flat, for which Schumann always hankers). The next
quatrain is based entirely on a punning use of the chord on the
flattened sixth of the scale, alternating with the dominant
seventh first in the key of B flat, then of E flat (Ex. 16), and this,
significantly, is the heart of the poem's mystery, which lies in the
words 'was ich erschau' in deinem Grund ist nicht Gewöhnlichen
zu nennen' ('what I descry in your bowl is not to be spoken of
before the common herd'). After this the last quatrain is devoted
to depicting the slow movement of the moonlight down the
valley and the solemn tolling of midnight. This passage is
another example of Schumann's chains of dominant sevenths

EX.16

and very effective it is here, with the voice released at last from
its note-by-note bondage to the piano and even varying slightly
the ♩ | ♩. ♪♪ ♩♩ | ♩. ♪♩ rhythm which Schumann pursues
inexorably throughout the whole song.

   Schumann himself was quite aware of the novelty and
originality of many things in his songs. On 31 May 1840 he
wrote to Clara that 'it often seems to me as though I were
treading quite new paths in music' ('als käme ich auf ganz neue
Wege in der Musik') and a year later we find him writing to
Kossmaly, author of an essay on 'The Song', with proper
consciousness of his own originality:

> It vexed me rather that you put me in the second class. I did not expect to
> be included in the first class, but I believe I can claim a place of my own and
> least of all am I pleased to be ranked with Reissiger, Curschmann, &c. I
> know that my endeavours and my means far exceed theirs, and I hope you
> will remember that and not call me vain, which I am far from being.

He was generous in proclaiming the influence of Clara on the
compositions of 1840. 'When I was composing them,' he writes
to her, 'I was entirely taken up with you. It is not possible to
write such music without a bride such as you—and I mean that
as high praise of you.' But the desirability of Clara's influence on
the final versions of some of the songs has been questioned. Not
her personal influence on Schumann, who told the same
Kossmaly that his music had always been 'the expression of
himself both as human being and as musician', with no cleavage
or hostility between the two. But Clara Wieck was first and
foremost a pianist formed in the classical and neo-classical
school, with none of the instinctive leanings to Romanticism

which her husband had always felt. Her musical taste leaned towards Mendelssohn quite as much as towards Schumann and there is, to say the least of it, a strong tradition (traceable directly to Theodor Kirchner, a disciple and friend of Schumann) that she persuaded Schumann to make various alterations in the text of his songs, always in the interest of academic 'correctness'. Schumann, in his aphorisms, expressed himself quite un-ambiguously on the subject of revisions: 'Two different readings of the same work are often equally good' (*Eusebius*), 'The original one is generally the better' (*Raro*). And in point of fact the alterations that we can trace from his notebooks are seldom of great importance and more often than not aim at perfecting the prosody of a song rather than any specifically musical improvement.

It was Schumann's practice to go on making alterations, which were often continued to the last moment, when the music was in proof. It will only be possible here to give a few typical examples of Schumann's revisions, as any more thorough study

EX.17

would necessitate a prohibitive amount of quotation. Most interesting of all is to see[3] how in many cases the songs which seem the most spontaneous and 'inevitable' were actually those on which Schumann took the longest time to make up his mind. Thus there exist three versions of 'Mondnacht' (Op. 39) (Schumann gave his mother-in-law a manuscript copy which differs from both the original and the corrected, published version), in which the composer seems chiefly concerned with giving the text and his illustrations more perfect expressiveness, though the musical character of the song receives hardly any alteration at all. On the other hand 'Frühlingsnacht' has a major alteration of the text of bars 15–17, as important as some of the major operations to which Beethoven's sketch-books bear witness. Compare the simplicity and expressiveness, the apparent 'inevitability' of the published version with the fussiness and ineffectiveness of the original (Ex. 17), in which the word 'herein' is repeated for no reason.

On the other hand, was Clara responsible for the 'toning down' of bars 69–76 of 'Das ist ein Flöten und Geigen', in which the chords of the original three-bar version (Ex. 18) seem to suggest either the climax of the festivity or the dagger-thrusts in the unfortunate lover's heart—or, knowing Schumann, both? Certainly the contradictory spirit of Jean Paul hangs over the whole song (Schumann probably had the last scene of the 'Flegeljahre' in mind, the ball and its extraordinary results for Walt, Vult, and Wina), for over bars 75–8, which show the lover moving slowly away from the scene of his misery, Schumann wrote 'Vivat hoch' which, presumably on reflection, he enclosed in brackets. Dr Wolff sagely remarks that this

EX.18

---

[3] Viktor Ernst Wolff: *R. Schumanns Lieder in ersten und späteren Fassungen* (Leipzig, 1914). Wolff's 'first' versions, however, are themselves fair copies, not initial sketches.

outburst is a 'memorial to Schumann's Janus-like temperament. In the heat of creation Florestan was led by his dramatic visions to make such annotations, which Eusebius later bracketed in a moment of cooler reflection.'

The manuscripts of the *Dichterliebe* are particularly heavily scored with revisions. Here again the most apparently spontaneous songs—'Die Rose, die Lilie' and 'Hör' ich das Liedchen klingen'—are shown to be the fruit of considerable reflection, if only on points of prosody, while 'Am leuchtenden Sommermorgen' appears as a minutely calculated work of art, each of whose effects is carefully planned and either subtly prepared or deliberately surprising (e.g. the inclusion of two bars in the piano part between the first two lines of the poem, bars 6 and 7, in which the harmonic effect of bars 8–9 is cunningly prepared). Schumann did not immediately make up his mind on the dramatic enharmonic modulation at the end of 'Die alten, bösen Lieder', where (bar 49) the effect of the dissonance was originally intensified by two quavers (E natural followed by C sharp) on the 'und', later replaced by the less violent single crotchet C sharp. Was this Clara's handiwork? Certainly I should suspect her influence in the change (bars 59–60) from the unconventional quintuplets in the manner of Chopin to the conventional gruppetti before the sixth beat in each bar.

Sometimes Schumann's second thoughts seem to have been definitely disimprovements. Why, for example, did he alter the opening bars of 'Liebesbotschaft', in which the original version opens canonically, with one bar of piano solo (bar 1 of the final version) and the voice following with the same melody one bar later? Clara cannot have objected to anything so academically respectable as a canonic imitation. Internal evidence provided by Schumann's very rare use of canon in a few other songs would suggest that he regarded this device—as we have seen that he regarded chromaticism—as producing a definite effect (presumably of ancient, semi-ecclesiastical venerableness, as in 'Auf einer Burg' (Op. 39)) and this was certainly out of place in setting the amorous platitudes of Reinick's 'Liebesbotschaft', though from the purely musical point of view the song loses interest in the final version. Schumann would perhaps reply that what concerned him first and foremost was old Wiedebein's advice, to 'look to truth above all'.

Liszt praised Schumann for choosing as texts for his songs poems 'whose beauty of form proceeded from a feeling capable of a still higher expression than words could give'. Actually the form of the poem was the only thing about it which Schumann tended to treat with little respect, repeating to his heart's content if he felt musically inclined, as we have seen. But Liszt was right in praising Schumann's sense of musical possibilities in a poem. There are very few instances of his setting to music poems complete in themselves and lacking in the atmospheric suggestion which was the first necessity in Schumann's case. The five poems by Mary, Queen of Scots (Op. 135) are certainly one example. In translation they lose the pathos of their sincerity and verbal simplicity (especially No. 5, 'Gebet', whose whole charm lies in its tripping Latin assonances); and it was probably the glamour of the author, unwittingly transferred to her poems by Schumann (as in the case of Elisabeth Kulmann), which made him choose such completely un-romantic words. But in general we have to go forward to Hugo Wolf before we find another German song-writer with such a fine instinct in choosing and setting his poems. Brahms was a more perfect craftsman than Schumann, with a more genuine understanding and less sentimentality in his approach to German folk-song. He was rhythmically far more adventurous, but at the same time he knew when the listener would have had enough of a good thing and he could never have written a whole song, like Schumann's 'Abendlied', with cross-rhythms persisting throughout. On the other hand his north German tendency to sentimental brooding led him to choose poems such as Klaus Groth's 'O wüsst ich doch den Weg zurück', Daumer's 'Wir wandelten' and Allmers' 'Feldein-samkeit' for some of what proved to be his most successful songs; and the quality of the poetry had a definite influence on the quality of the music. Schubert might have set such poems with impunity; and Schumann might have conceivably chosen them, though they would have become poor songs in the second or third rank of his output. Brahms's masterpieces—for master-pieces they are, in their way—are founded on banal literary material and emotion, which they faithfully reproduce with great charm. Whereas the great songs of Schumann are a joint triumph for Schumann and Heine, or Eichendorff, many of the great songs of Brahms are a triumph (sometimes Pyrrhic) for

Brahms alone. It is only in the folk-song pastiche, where Schumann generally failed so signally, that Brahms is consistently his superior. We have only to compare the older man's 'Liebhabers Ständchen' or 'Unterm Fenster' with his successor's 'Vergebliches Ständchen' (Op. 84); 'Lied eines Schmiedes' (Op. 90) with 'Der Schmied' (Op. 19); 'Sonntags am Rhein' with 'Sonntag' (Op. 47) (and the list could be prolonged) to see Brahms's unquestioned superiority in this vein.

Schumann's influence on Liszt as a song-writer was, I rather think, unfortunate. In his settings of Heine, Liszt owes nothing to Schumann; he was a dramatic miniaturist, as we have seen, and a lyrical poem like 'Du bist wie eine Blume' appears as an exception. How differently the two men treated even the same poems of Heine's can be seen by comparing their settings of 'Anfangs wollt' ich fast verzagen' and 'Im Rhein, im heiligen Strome' (which Liszt titled 'Im Rhein, im schönen Strome'), where Liszt paints and dramatises phrase by phrase, bar by bar, and Schumann lays down a solid rhythmic pattern which is pursued almost owlishly from beginning to end. It was in his settings of Goethe that I suspect Liszt of following the bad example of Schumann, in wandering from section to section and hammering together what often amounts to four sides, a lid and a bottom all belonging to different boxes. The chronology of Liszt's songs is uncertain, even such a trifle as 'Dichter, was Liebe sei' appearing in three different versions dating from 1844, 1855, and 1878. But the majority were written in the 1840s, after Liszt settled at Weimar (1842). And certainly a big blowsy song like 'Ich möchte hingehn' or the sprawling settings of 'Wer nie sein Brot mit Tränen ass' (1845 and 1860) were written with a full knowledge of Schumann's songs, good and bad. That Liszt did not always distinguish the one from the other is shown in his choice for piano transcription. Thus by the side of 'Widmung' and 'Frühlingsnacht' stand thoroughly second-rate songs such as 'An den Sonnenschein' (Op. 36) and (significantly) 'An die Türen will ich schleichen' (Op. 98a) and the rather colourless numbers from the *Liederalbum für die Jugend*. Perhaps the truth is that both Liszt and Schumann suffered, especially as they grew older, from the same, typically Romantic failing—a passion for the vague, large and high-sounding which led both men in their different ways to a form of musical meandering among windy generalities.

Hugo Wolf, who owed much to Schumann, instinctively turned to songs of an entirely different kind. A dramatic miniaturist like Liszt, he developed and made explicit the drama which is often only hinted at by Schumann. The two men's settings of Mörike's 'Er ist's' and 'Das verlassene Mägdlein' provide typical examples of their different methods. Schumann's setting of 'Er ist's' is included in the *Liederalbum für die Jugend* and it has the self-conscious simplicity which, as we have seen, Schumann associated with children. It is marked 'innig', whereas Wolf's mood is clearly shown by 'sehr lebhaft, jubelnd' ('very lively, with jubilation'). Both songs open with two bars for the piano, Schumann's based on a descending scale passage with dotted rhythm, bringing in the voice on his favourite dominant seventh in the key of the dominant (dominant seventh of E in the key of A major) (cf. 'Loreley' (Op. 53) and 'Schöne Fremde'). Wolf merely sets in motion a vigorous rhythmic figure which is to dominate the whole song, a broken tonic triad (G major) over which the voice enters with an opening phrase spread over the notes of the triad. At the end of the first quatrain, where Schumann has modulated through the minor of the mediant (C sharp) to a conventional close in the key of the dominant in the interlude for the piano, Wolf has quickly reached the remote key of C sharp major. A comparison of the two settings of the next two lines, with exactly the same rhythm in the vocal parts (Ex. 19, *a* and *b*) will show how completely

EX.19

a) Schumann

Veil-chen träu-men schon   wol - len bal-de kom-men   Horch!

b) Wolf

differently two composers can envisage the same poem. The piano part is of primary importance in each case, of course. Schumann's is actually the more independent, Wolf's more in the nature of an accompaniment. Both composers instinctively aim at expressing the atmosphere of expectancy, which is the note of the poem, by a series of generally unresolved dominant sevenths, Schumann's naturally far less sophisticated than

Wolf's. Both emphasize the 'Horch!' by a sharped dominant seventh right out of the key of the preceding bars: but Schumann continues the shy, hesitant ('etwas zurückhaltend') manner and the rather self-conscious, Dresden-china prettiness already suggested by the mordents in the piano part, whereas Wolf quickly works up to his climax and gives final vent to his excitement in a long piano postlude (twenty-one bars in a song of under sixty). The last lines of the poem read in Wolf's version:

> Frühling, ja, du bist's!
> Frühling, ja, du bist's!
> Dich hab' ich vernommen!
> Ja, du bist's!

This already has two unwarranted repetitions. Schumann, in a happy haze, 'innig' rather than 'jubelnd', murmurs the words again and again:

> Frühling, ja, du bist's! ja, du bist's, du bist's!
> Dich hab' ich vernommen, ja, du bist's!
> Dich hab' ich vernommen,
> Frühling, ja, du bist's! ja, du bist's!
> Ja, du bist's, du bist's, du bist's,
> Dich hab' ich vernommen,
> Ja, du bist's!

There is only the shortest postlude (in which the 'pretty' mordents recur). These are the dangers of 'Innigkeit' and too much Jean Paul, the dreaming aloud and the soulfulness which is not so simple and unselfconscious as it would like to seem, but pins on some arch little ornaments and is not above an occasional simper.

Between the two settings of 'Das verlassene Mägdlein' the resemblance is so marked that it is impossible that Wolf had not got Schumann's in mind when he wrote his own.[4] But

---

[4] Since writing this passage and what follows I have found confirmation of this in Ernest Newman's *Hugo Wolf*, p. 182, n. 3, where he says that in a letter to Eckstein dated 27 March 1888, Wolf wrote that he greatly admired Schumann's setting of 'Das verlassene Mägdlein' and had no intention of setting the poem himself: but that being deeply affected by the poem the music came to him almost against his will. The comparison of the two settings which follows was written without knowledge of Mr Newman's (p. 190), with which I am in general, but not complete, agreement, as comparisons will show.

Schumann's is a *cri de cœur* whereas Wolf's is a picture. Schumann cares nothing for the suggestion of early morning cold and desolation which Wolf gives in his piano prelude, still less for the sparks which Wolf strikes as the girl lights the fire, interrupting the heavy tread of the rhythm ♩ ♫♩ ♫ , an exaggeration of Schumann's, which is alleviated by the movement of the inner parts. Schumann places his two-bar piano interlude between the two lines of the poem where the girl lights the fire and then sits gazing miserably into it. Wolf, with a better literary and dramatic instinct, puts his four bars of interlude after the lines and makes them depict the girl's misery, so that his 'plötzlich, da kommt es mir' sounds like a real, sudden memory. He paints the ecstasy of the girl's dream of her lover in three repeated phrases, ritardando, and ending on a pause before the tears start falling hopelessly. Schumann, aware only of the girl's misery, carries straight on from the dream to the tears with an accentuation of the harmonic clash (dominant minor ninth, Ex. 20; cf. 'Mignons Lied' (Op. 79)) which is a characteristic feature of the whole song. Wolf, in an era of greater harmonic sophistication, when minor ninths had become small beer, chooses a subtler chord as the symbol of the girl's grief, the augmented triad or 'whole-tone' chord in its various positions. Again Wolf's postlude, with its bare open fifths, suggests the misery and desolation of the opening, whereas Schumann has no postlude at all but comes to rest on a rather inept 'tierce de Picardie', the tonic triad with the major third instead of the minor. The unusually chromatic harmony in Schumann's setting (in accordance with his regular association

EX.20

Trä - ne auf Trä- ne dann stür - zet her - nie - der

of chromaticism with extreme grief) becomes with Wolf simply a pianissimo dying-away of open fifths, which accords well with the simple emotion of Mörike's poem and suggests the girl's desolate hopelessness. Not that a comparison of the two songs need result in a value judgement between them; it is merely instructive to see how differently the two composers work on a given poem which admits (unlike 'Er ist's') of only one emotional approach, from what different angles they approach it and what are the alterations Wolf makes in what is almost certainly an (unconscious) 'rethinking' of Schumann's music as well as Mörike's verses.

There are other songs by Wolf which often seem to be sophisticated rethinkings of Schumann's material. The two 'Zigeunerliedchen' from the *Liederalbum für die Jugend* have already been mentioned and a comparison between the first of them ('Unter die Soldaten'), Ex. 10, with, say, the opening of Wolf's 'Schweig einmal still' (*Italienisches Liederbuch*, Vol. 3 No. 43) will give an idea of what I mean. Wolf's soldier songs, also, seem to hark back to Schumann's (compare, for example, Schumann's 'Die Soldatenbraut' (Op. 64) with Wolf's 'Der Tambour') though again the later songs have a variety, both rhythmic and harmonic, and a dramatic sophistication which often make the earlier seem homely and naïve in comparison.

In Schumann's special domains he is supreme and need fear no comparison: the romantic night-pieces like 'Der Nussbaum', 'Frühlingsnacht', 'Mondnacht', 'Schöne Fremde' or what may for want of a better title be called psychological, rather than dramatic lyrics as those of the *Dichterliebe*, 'Wer machte dich so krank?', 'Die Fensterscheibe'. In the night-pieces he found musical expression for a specifically new, romantic emotion. (New to music, that is to say: Novalis's 'Hymnen an die Nacht' were nearly fifty years old in 1840, while Edward Young, whose *Night Thoughts* originated the cult, had died as long ago as 1765). In Schubert's 'Nacht und Träume' there still remains something of the eighteenth century, a marmoreal serenity and an absence of the Orphic or Dionysian quality which begins to appear in Schumann's night-pieces ('In der Nacht', from the *Fantasiestücke* for piano, for example, as well as the songs) and found its fullest expression in that 'vast nocturne', Act II of *Tristan*. It is a specifically German quality, quite unlike the elegant melancholy of Chopin's Nocturnes or Tom Moore's 'At the mid hour of night', no less beautiful for being tributes to a literary fashion but only incidentally night-pieces; whereas Schumann's music really suggests another world of feeling, solemn and mysterious, peopled with poetic imaginations which would fade in the light of common day, and foreshadow Tristan's 'göttlich ew'ges Urvergessen'.

To speak of Schumann as merely a miniaturist is a by now dangerous platitude; but it is certainly true of Schumann as a song-writer. He managed his larger songs worse than any of the great German song-writers, Schubert, Brahms, or Wolf; but in return he showed himself a master of pregnancy and compression in the psychological lyric, a genre which is virtually his invention. He was not a dramatist like Wolf; we have seen how differently the two men reacted to the same poem. But Schumann's literary background and his reading had given him an instinctive understanding, a psychological acumen, which combined with his lyrical gift and his own intensely emotional character to create something like a new form. While much of Schumann's music, including a considerable proportion of his songs, bears the traces of the naïvely emotional and complacent background which determined the composer's mentality, there remains an irreducible minimum of works which have that ageless quality which is the only certain hallmark of genius. At the very centre of that core are the great songs, so that if future generations remember Schumann for nothing else, he can hardly fail to be counted among the very greatest of the German song-writers.

15

CLAIRE CROIZA[*]

In 1924 France celebrated the four-hundredth anniversary of the poet Ronsard's birth and during those celebrations Paul Valéry heard a voice, a speaking voice, that was new to him and impressed him profoundly. The voice was Claire Croiza's and Valéry in his enthusiasm spoke of it as 'the most expressive voice' (*la voix la plus sensible*) 'of our generation'. No poet, perhaps, could claim to possess a more expert understanding or a more exquisite natural sense of words than Valéry—their colour, texture, shape, association, and the vast variety of their mutual relationships, musical and intellectual. If it was the voice of a singer rather than an actor that won Valéry's supreme approval, this is much less strange in France than it would be in England. A correct and beautiful pronunciation and enunciation of the French language is an important part of the French singer's art. Much of the greatest French vocal music has been a kind of heightened declamation and the relationship between words and music has been closer in France than in any other European country. Thus Paris ousted Italian and insisted on truly French opera, with a French text, a hundred years before any other country; and the architect of that French opera, Lully, advised his singers to model their singing on the declamation of a famous actress of the day, La Champmeslé.

This strong literary affinity was at the base of Claire Croiza's art as a singer, and her mastery of the spoken word was the foundation of that understanding and mastery of verbal beauty on which Croiza's whole art as a singer was built.

It is therefore doubly surprising to find that neither of her parents was French. Her father was an Irishman and her mother an Italian, and her real name was Conelly. She was born in 1882, studied with Revello and had a few lessons with Jean de

[*] Martin Cooper, *Ideas and Music* (London, 1965), pp. 199–203.

Reszke; and her first engagement was at the Nancy Opera House in 1905. She made her first appearance in Paris the following year, when she sang the Angel in *Gerontius*—somehow an improbable part but one which made her immediately remarked upon. Her first great successes came to her between 1908 and 1912, when she was at the Brussels Opera. Her mezzo-soprano voice belongs to a category which has not been very generously treated by composers of opera. However, in those Brussels years she sang the part of Dido in Berlioz's *Les Troyens*, the title roles in Massenet's *Thérèse* and *Ariane* and Charlotte in *Werther*, Saint-Saëns's *Dalila* and Bréville's *Eros Vainqueur*. Most important of all, in 1909 she sang Gluck for the first time—*Orfeo* and the part of Clytemnestra in *Iphigénie en Aulide*. After the War her appearances in opera became increasingly rare. Instead she devoted herself more to concert singing and particularly to the performance of French songs.

That she was now at the height of her powers, is shown by Valéry's praise. It was during these years, too, that she first consciously developed the literary side of her art, arranging with Jacques Copeau a series of concerts under the title 'Poésie et Musique'. In these programmes Croiza recited and sang a group of poems chosen either from a period or from the works of a single poet—the Romantics, for instance, or Verlaine. She had long been an admired friend of the older generation of French composers, including Fauré, d'Indy, Saint-Saëns, and Debussy. Now she made personal friends among the musicians of a later generation, Caplet and Honegger, or those whose names were then less well known to the public—Florent Schmitt, Déodat de Séverac, Albert Roussel, and Pierre de Bréville. Her championship of these composers' songs, along with the already 'classical' songs of Fauré, Duparc, Debussy, Chausson, and Ravel, did more than any other form of publicity to make them widely known.

Croiza's voice was the perfect instrument for the interpretation of French song. The mezzo-soprano has neither the brilliance of the soprano nor the often dangerous richness of the contralto. It is the viola among voices, impressing by its distinctive colouring and by art rather than by nature. Croiza's tone was firm but never rich. There is a certain spareness which enhances the beauty of her vocal line, a sobriety of colouring and an instinctive avoidance of dynamic extremes, all of which are characteristically French and allow the listener to become aware of her perfect enunciation and the subtlety with which she characterizes each phrase.

There was nothing that Croiza disliked more than false pathos. She was fond of Debussy's description of French music as 'quelque chose comme la fantaisie dans la sensibilité', but deplored the fact that she found among her pupils so much sensibility and so little fantasy. Gaiety, wit, fantasy, irony, were the characteristics of her art as of her conversation, and she deplored the rather mawkish melancholy particularly common, and perhaps secretly cultivated, among her women pupils. 'If there is a moment, even a second, in a song when the sense allows you to smile, smile for heaven's sake,' she once said. She had excellent advice for those who were studying a song such as Duparc's 'Chanson Triste' which can easily be sentimentalized. First, complete sincerity—'the singer must strip to the soul, as the artist's model strips to the skin. It's not easy the first time, but it must be done at all costs—give everything and then one's interpretation gains life and one's tone emotional depth.' Secondly, physical vitality, which will ensure the singer against sagging rhythms and flabbily emotional phrasing. 'Vous avez une belle voix, madame,' she said to a surprised pupil who had sung 'Chanson Triste', 'chaleurisez-la par votre chaleur animale, par votre force animale.'

The first secret of interpretation according to Croiza was self-forgetfulness. This, as she knew, is an art in itself and one which needs long practice to achieve. 'Your expression', she would say, 'will only be good if you do not try to be "expressive". Never give anything more importance than it has by right, or you will ruin everything'—this speaking of Poulenc's *Le Bestiaire*, her own singing of which is a perfect example of her verbal as well as her vocal skill. 'In singing', she added, 'one may prefer either the sound or the word. No sound however beautiful will ever give me personally the joy that I get from a beautifully enunciated vowel.'

She was a fully formed artist and no longer a young woman when she started singing the music of Milhaud, Honegger, Schmitt, and Roussel, and she found some of it hard on her voice and almost impossible to carry off with the poise that she considered essential to all singing. 'It's as though someone said to you: "Throw yourself from a fourth floor window and mind you fall gracefully",' she observed of one of Schmitt's works.

Although Croiza did not generally discuss songs in technical detail, she was a fanatical observer of the composer's intentions, a stickler for the exact observance of note-values and rests and above all of a strict and steady rhythm. After her performance in Fauré's *Pénélope* the composer is said to have remarked: 'How nice for once to hear my music as I wrote it.' If you follow her recording of Debussy's 'Il pleure dans mon cœur' from the *Ariettes oubliées*, for example, you will find that she takes no liberties whatever. And yet her performance seems to be perfectly spontaneous. Croiza knew her songs so intimately and so accurately that a textually flawless performance came to her naturally.

During the last twenty-five years of her life Croiza did an enormous amount of teaching, but at heart she remained a woman of the theatre. She felt very strongly about operatic production and once said that, if she had the money, she would like to tour the world producing four operas—*Carmen*, *Louise*, *Pelléas et Mélisande*, and *Pénélope*. Even in the concert-hall she insisted on the importance of stance, *tenue* (she herself always sang in a hat), and above all of facial expression. Too many gestures and movements she considered a mistake, but the face could and should aid and increase the expression of the voice.

She had the instinctive French fineness of perception, distinguishing nicely between categories. Of the Villon-Debussy 'Ballade' for instance, she insisted that it was a prayer, not a poem; and that no prayer must ever sound sad. 'A sad prayer, I think, must bore Le Bon Dieu'—and her own recording, one of the last she made, is characterized by a delightfully matter-of-fact tone of voice, which yet contrives to increase the impression of sincerity and urgency.

Claire Croiza died in May 1946 of a cerebral tumour. She had continued her teaching in Paris during the War and, with few foreign pupils and no foreign tours, her life was a hard one. There is no truth in the story at one time current that she in any way collaborated with the occupying Germans. When she was suspended from teaching at the Conservatoire during the first hysterical days of witch-hunting after the liberation of Paris, a group of writers and musicians signed a document asserting her innocence and recalling her lifelong work for French music at home and abroad. The signatures included those of Valéry, Marcel, Lacretelle, Désormière, Auric, Bréville, Poulenc, Honegger, and Münch. Her name was cleared, but nothing could replace the loss caused by her death which followed soon after. The extraordinary haunting quality of her voice, the somehow fragile and essentially feminine nobility of her musical nature, the aristocratic reserve allied with a deeply human pathos and simplicity, the gleaming wit and the dash of irony— these made up a unique artistic character.

16

PAULINE VIARDOT[*]

On whom was the character of George Sand's Consuelo based? Who sang the first performance of Brahms's *Alto Rhapsody*? Who sang Leonore in the first performance of *Fidelio* in Paris, and Azucena in the first London performance of *Il Trovatore*? To whom was Schumann's *Liederkreis*, Op. 24, dedicated? Whom did Berlioz originally want to sing Dido in *Les Troyens*? Who sang the title role in the spectacularly successful Paris revival of Gluck's *Orfeo* in the 1860s? Who introduced Rimsky-Korsakov's songs to Paris in the 1870s?

The quiz could easily be extended without altering the answer, which is in every case Pauline Viardot-Garcia, who was born in 1821 and died in 1910, the sister of 'La Malibran' and of Manuel Garcia, the great singing teacher. There can never have been a better subject for a biography, and it is difficult to imagine why no one has anticipated April Fitzlyon's *The Price of Genius*.[1] It may be because the world in general remembers this remarkable woman not for her musical gifts—and she was a pupil of Liszt, admired as a pianist by Clara Schumann and Saint-Saëns long after she had become a professional singer—but as the woman to whom the novelist Turgenev devoted his whole life. In this capacity she has been much criticized, even reviled, by Russian literary historians, and it is one of Mrs Fitzlyon's achievements to have cleared her name of the charges of enticing Turgenev from Russia and keeping him as a tame but sexually famished 'lion' in her family circle.

It is perfectly clear from this book that Turgenev was only one of many profoundly feminine men (others were de Musset, Gounod, Ary Scheffer, and, in a moment of pathetic emotional

[*] The *Daily Telegraph* 23 January 1965.

[1] April Fitzlyon: *The Price of Genius—a life of Pauline Viardot* (London, 1964).

weakness, Berlioz) who enjoyed the domination of this brilliantly gifted woman whose chief interests in life were professional and artistic rather than personal. Turgenev's role in her life was pathetic and undignified (when he was in Russia and she in France, he asked for her nail parings, and it is perhaps to her discredit that she sent them) but it was only when both were middle-aged that she openly accepted his homage and returned his affection. No conclusive evidence exists to show whether Claudie Viardot was, in fact, Turgenev's child, or even that the couple were lovers, though this seems probable.

Mrs Fitzlyon documents very fully George Sand's influence over the young, ugly, but enormously gifted Pauline Garcia. It was George Sand who warned her of the 'price of genius' and, with extraordinary far-sightedness, married her to Paul Viardot, a steady and prosperous liberal publicist and a passionate sportsman who adored her and was rewarded with dutiful affection and respect. The friendship between Turgenev and Viardot may be interpreted as a striking example of either the power of civilization over instinct or, alternatively, a profound cynicism. I should suspect, although Mrs Fitzlyon does not suggest it, that Viardot liked Turgenev personally and was not stirred to any intolerable jealousy by an admirer who showed so few specifically masculine traits of character.

Pauline Viardot's voice was a mezzo-soprano which she enlarged (according to *Grove*) to cover a span from C to F'''. The voice itself was by no means uniformly beautiful in tone and the singer was an ugly woman, a fact which coupled with her husband's political opinions, explains her difficulty in pleasing the Paris public. In St Petersburg and London during the 1840s and 1850s, on the other hand, and later in Germany, she was enormously admired for the musicality and dramatic power of her performances in a fantastic variety of roles. From Rossini's Rosina, Cenerentola, and Desdemona she passed to Bellini's Norma, Amina, and Romeo and Donizetti's Lucia and Leonora, while her classical roles included Zerlina and Donna Anna as well as Orfeo and Alceste. Probably the greatest success of her whole career was the part of Fidès in *Le Prophète* and her Meyerbeer roles included both the soprano parts in *Robert le Diable* as well as Valentine in *Les Huguenots*. It is an interesting fact that Berlioz, who wanted her to sing Dido, always criticized her for indulging in excessive ornamentation, a fault more characteristic of the brainless canary than the serious artist.

Pauline Viardot was, in fact, an unrepentant musical conservative, who had no use for Wagner's music, although in 1860 she had the extraordinary experience of sight-reading the second act of *Tristan* with the composer in her own drawing-room, when she found his tempos culpably unpredictable. Although she admired *Die Meistersinger*, probably the most advanced music that she spontaneously appreciated was that of Chopin, whom she knew personally, and it is an extraordinary tribute to her artistic powers that he approved of her vocal versions of some of his mazurkas. When she might have been singing Wagner, she was creating the title-role in Gounod's *Sapho* and Massenet's *Marie-Magdeleine* or thrilling audiences with her performances as Rachel in Halévy's *La Juive*; but she did much to further the cause of Russian music in France and her *Orfeo* performances revived interest in Gluck which had almost vanished.

Mrs Fitzlyon sums up Pauline Viardot's character very shrewdly and in fact accepts George Sand's estimate of her, written in 1852. She spoke there of 'the profound egoism of the first-class artist' and of 'a highly-developed instinct for self-preservation'. Pauline Viardot, she says, 'protected herself instinctively against possible worries, complications, even emotions. She was of the school which prefers no feelings to unpleasant ones and will sacrifice positive happiness, if it carries with it the risk of possible unhappiness.' It was Turgenev's tragedy that he was made of less stern stuff, but had Pauline Viardot been a gentler and more feminine woman, he would never have felt for her as he did.

PART V   ON THE AIR: CONTRIBUTIONS TO THE BBC

17

THREE BROADCAST OPERAS

Les Dialogues des Carmélites*

On 7 July 1792, the Théâtre Feydeau in Paris gave the first performance of an opera by Devienne entitled *Les Visitandines*— or *The Nuns of the Visitation*—with a libretto by L. B. Picard. The Visitation nuns were—and still are—a religious order much concerned with education, and the plot of Devienne's piece (which was in fact an *opéra comique*) shows a young blood and his valet disguising themselves in order to obtain entrance into a convent where the heroine is a novice. Much of the dialogue is equivocal in character and contains a number of rather primitive and, to our thinking, mild blasphemies which, at the height of the Revolution, accounted for a good deal of the work's success. They also explain the title of a parody which appeared anonymously the following year—*Les Putins cloîtrées* or *Tarts in the convent*. During the next twenty years *Les Visitandines* was given in translation—and in many cases disguise—in Germany, Switzerland, Russia, Holland, Sweden, Copenhagen, Madrid, and Dublin—but not apparently in London. The anti-clericalism of Picard's libretto was given a different form in each country, no doubt, but always formed the chief attraction of the work.

It is at first difficult to see how the life of religious orders, whether in convent or monastery, has anything to offer to the playwright or opera composer. Prayer and contemplation; education; nursing and the care of the poor or afflicted; intellectual studies—these all form a barren prospect for the theatre. Drama enters such lives as these chiefly in the form of failure, sometimes tragic but more often pathetic; and so the monks and nuns we see in the theatre or in literature are for the

<hr>

* BBC: transmission date unknown.

most part caricatures—gross contradictions of their pretensions, like the nuns in Diderot's *La Religieuse* and Manzoni's *I Promessi Sposi*—or colourless paternal figures like Friar Lawrence in Shakespeare's *Romeo and Juliet*.

The monks and nuns of the operatic stage fall into the same two classes. On the one hand we have Mozart's and Rossini's venal and malicious Don Basilio (not in fact a monk, but a secular priest), the mock-nuns in Rossini's *Le Comte Ory*, the Grand Inquisitor in Verdi's *Don Carlos*, the drunken begging-monks in Mussorgsky's *Boris Godunov*, the fanatic Dosifey in his *Khovanshchina*, the neurotic Renata of Prokofiev's *The Fiery Angel*, and the gross monks of his *The Betrothal in a Monastery*—and that is by no means the total. On the other hand, the list of the virtuous is small and unimpressive—hardly more than the Hermit in *Der Freischütz*, Friar Lawrence in Gounod's *Roméo et Juliette* and perhaps we should include the Cardinal de Brogni in Halévy's *La Juive*.

In the past, public opinion in Protestant countries has not wanted to see on the stage any representatives of the religious life except those which disgrace it; and in Catholic countries it was felt, at least until very recently, that monastic life is of its essence undramatic unless disrupted by interior frictions any-way too petty to be worthy of the stage or by some exterior disaster. Monks and nuns who have somehow failed in their vocation, or never had a true vocation in the first place, make the same sort of news as failed marriages. Happy vocations, like happy marriages, have no obvious dramatic interest.

It is only in the present century that any serious attempt to portray the monastic life as such—rather than individual and mostly 'rogue' religious outside their monastic settings—has been made in the opera-house. Massenet was the first to present a good-humoured, slightly cynical but not impossible picture of life in a monastery. This was in *Le Jongleur de Notre Dame*, which shows a community containing the familiar medieval types of a wine-loving Brother Cellarer and a devoted artist-illuminator, as well as the juggler whose only offering to God consists of his handstands. That was in 1902. Sixteen years later Puccini included in his operatic triptych *Suor Angelica* which presents a faithful and touching, if slightly monotonous and frankly superficial picture of life in a convent. But even here the drama is provided by the young nun who has been sent to the convent by her family after bearing an illegitimate child and by the subsequent visit of her aunt, whose character seems in some ways to be a preparation for the coldness and cruelty of the later Turandot.

Francis Poulenc came to the libretto of his *Les Dialogues des Carmélites* in a roundabout way. The original story is by the German writer Gertrud von le Fort and is called *Die letzte am Schafott* (Last on the Scaffold). This was used as the basis for a French play entitled *Les Dialogues des Carmélites*, by Georges Bernanos; and also for a film scenario—and it is from these two sources that Poulenc's libretto is drawn. Gertrud von le Fort's story, which is based on a historical incident during the Terror years of the French Revolution—the eviction and execution of the Carmelite nuns of Compiègne in the autumn of 1793— shows first the family background and then the convent life of a young noblewoman who enters the Carmelite order and is eventually guillotined with the rest of her community. Highly strung and precariously balanced emotionally, Blanche de la Force has already had a frightening experience of mob violence before she enters the community, and she is haunted by the fear of her own physical cowardice and her total inability, as she believes, to face the possibility of martyrdom. At first she sees the convent as an escape, a hiding place for that nervous timidity which disgraces her in her own eyes and in those of her aristocratic family. This attitude is in fact a thinly disguised form of pride, as the Reverend Mother shows her in their first interview. Gertrud von le Fort, Bernanos, and Poulenc himself were all, in their very different ways, devoted and intellectual Catholics; and this interview shows both in its language and its attitudes a profound and sympathetic understanding of Catholic psychology. 'We are neither a school of mortification nor a conservatory of virtue,' says the Mother Superior, 'but a house of prayer. Those who do not believe in prayer can only regard us as impostors and parasites.' She warns Blanche that she must expect great trials. 'What matter,' replies the girl, 'if God gives me strength to bear them?' 'What God wishes to test in you, my child,' says the old nun, 'is not your strength but your weakness.' And the drama lies precisely in this testing—Blanche's initial failure to face danger when the community is disbanded,

her flight and triumphant rejoining of the community at the very
foot of the guillotine.

There are in fact two levels of existence in the opera: the easy,
conversational, well-bred world of Blanche's family in the
opening scenes and in a single scene of Act II; and the convent
world, in which genuine and profound religious aspirations are
repeatedly interrupted, or at least interlarded by the irruption of
minute psychological repercussions and the play of personal
sympathies or antipathies inseparable from any community life.
Poulenc is extraordinarily successful in finding musical equival-
ents for both; and to do this, he had recourse, very characteristically, to a number of totally dissimilar traditions from which he
formed, by both blending and adapting them, a musical
language that is wholly individual. *Les Dialogues des Carmélites*
is primarily a parlando opera interrupted by arioso passages and
ending with the purely melodic 'Salve Regina' chanted by the
community as they go to the guillotine—a wonderfully effective
musical 'release' after so much conversation. The sources on
which Poulenc drew for his parlando style were in the first place
Massenet—the Massenet of *Thérèse*, another opera with a
French Revolutionary setting—and Debussy, from whom he
learned the flexible and fluid setting of the minimally accented
French language. From Debussy, too, comes Poulenc's regular
division of the music into four-bar phrases in which the second
two repeat, or only slightly vary, the first two, a practice which
Debussy himself had imitated from the Russians.

Harmonically the musical style of the opera is almost
ruthlessly triadic and homophonic. The scenes of secular life
contain powerful memories of Massenet in the textures and
accompaniment figures; and also in the melody that dominates
Blanche's interview with her brother in Act II. In the convent
scenes Poulenc's own sophisticated and very sweet triadic
harmonies alternate with a much more austere, modal harmonic
language which, when added to the two-bar structures, often
make a distinctly Russian impression. But it is never long before
Poulenc returns to his own inimitable brand of discreetly
luscious harmonic progressions underpinning the conversational, parlando melodic line.

The main role, that of Blanche de la Force, is well characterized. Her timidity and her complex reaction to it are

presented in musical terms that suggest the inner strength which develops as soon as she learns to accept, rather than to fight against her weakness. Against the complications of Blanche's character, the total naturalness and simplicity of the country-bred Sister Constance is tellingly set in the high-spirited style of Poulenc's rural songs; and it is difficult not to remember the two contrasted sisters, Charlotte and Sophie, in Massenet's *Werther*.

The Mother Superior, whose death scene concludes the first act, is the most remarkable character in the whole drama; and her terror of death, which, as she knows, is so unedifying to the rest of the community, is beautifully described or suggested in the text, though less happily in the music. So that Sister Constance's instinctive understanding of Blanche's own terror, as something which the Mother Superior has suffered and conquered on her behalf, carries immediate conviction. But— and here is the major doubt of the opera as a whole—was Poulenc really able to give these fine points of psychology convincing dramatic expression or to find a musical style adequate to express these essentially interior events? Poulenc achieves, far better than Puccini in *Suor Angelica*, a music appropriate to the external course of convent life, uneventful but inspired by much deeper preoccupations and ideals. These, though clearly expressed in the text, hardly find expression in Poulenc's charming, easily-flowing music, with its many reminiscences of the operatic past. It is only in the final 'Salve Regina' that the heroism, that has been implicit in the community's refusal to abandon their vocations, finds explicit musical expression. To this objection Poulenc himself might answer that heroism in real life is generally no more than implicit until the moment when the ultimate test demands a show of its true strength.

Iphigénie en Tauride[*]

Iphigénie en Tauride was the last but one of Gluck's operas, and it was mostly written in Vienna during the spring and summer months of 1778. He worked concurrently on this and what proved to be his last opera, the very different *Echo et Narcisse*. Both were written for Paris where, under the patronage of his

[*] BBC 17 November 1973.

former pupil Marie-Antoinette, Gluck's operas were having an almost unparalleled success. When he returned to France in November, it was to be for his last visit. On 18 May 1779, *Iphigénie en Tauride* had its first performance and at the end of that October, Gluck went home to Vienna. At sixty-five his health was finally breaking down, and although he lingered on until 1787, his composing days were over.

Contemporary Parisian audiences were unanimous in their admiration of *Iphigénie en Tauride*, and their verdict has been wholeheartedly endorsed by posterity. If Gluck is to be judged by any single one of his operas—and there are some fifty of them—it is surely this; and in view of his operatic theories, it is not by chance that Guillard's libretto is the best 'opera-book' that ever came into Gluck's hands. It seems likely that the original idea of the work goes back to discussions between the composer and the librettist of his earlier *Iphigénie en Aulide*— Le Bland du Roullet; and it may even be that Guillard did no more than complete or polish a libretto already sketched by du Roullet.

The drama is based on that of Euripides, but the discrepancies between the two are almost as important as the similarities. It is a fascinating historical chance that, at exactly the same time as Gluck was completing his score, Goethe was preparing his first, prose version of the same story. This was to be still another interpretation of the journey of Orestes to rescue his sister Iphigenia from her mysterious, supernatural banishment in what we know as the Crimea; and so to fulfil the conditions imposed on him by Apollo in order to purge himself of the guilt of matricide. His haunting by the Furies—impersonations, as we may imagine them, of his profound sense of guilt—was used by Aeschylus as the theme of a whole play. Although it is not mentioned in Euripides, both Guillard and Goethe found its dramatic potentialities too tempting to resist; and the fourth scene of Gluck's second act—the first emotional crisis of the work—is a magnificent musico-dramatic picture of the visitation of Orestes by the Furies and the ghost of the murdered Clytemnestra. In *Orfeo*, Gluck had already attempted something similar; but this scene is musically much closer to a passage in his Italian opera *Telemacco*.

The character of Orestes, and his relationship to his companion

Pylades, are totally different in Euripides and Guillard. Euripides knows nothing of Guillard's (and indeed Goethe's) hag-ridden, neurotic Orestes; or of the high-pitched sentiments that lead to his dispute with Pylades as to which of them shall sacrifice his life for the other—a characteristic scene from French classical drama. The only note of self-pity in Euripides's Orestes is in his answer to Iphigenia's question whether Agamemnon has a son still living in Argos. 'He has,' says Orestes, 'a wretched one, nowhere and everywhere.' The corresponding scene in the opera is not the only occasion when the character Orestes, and his relationship with his sister Iphigenia, seem clearly to foreshadow Wagner's Siegmund and Sieglinde.

Iphigenia herself is given, in addition to the human dignity and woman's charm of Euripides's heroine, a modesty, a purity of intention and a piety that make her at least as Christian a figure as, say, Racine's Phèdre. This side of her character finds unforgettable expression in two of her arias—the hymnlike 'O toi qui prolongeas mes jours' in Act I and 'O malheureuse Iphigénie' in Act II, where the major mode and the oboe solo combine to present a picture of grief and despair heroically transcended. Another side of her character, more passionate, is shown in the big dramatic A major prayer that opens Act IV, another borrowing from the earlier *Telemacco*; and Gluck cleverly chose the simple, naïve manner of contemporary *opéra comique* for her 'D'une image hélas! trop cherie' and for the trio in Act III where she details the arrangements for the two victims' escape. Guillard's Thoas is a barbarian even more superstitious, bloodthirsty, and stupid than Euripides's—totally different from Goethe's magnanimous tyrant, and musically a close ancestor of Beethoven's villainous Pizarro.

From the very opening the orchestra plays an important role. The unbroken transition from the storm (another similarity to *Die Walküre*) to the chorus of priestesses and thence to Iphigenia's narration of her dream, was wholly original. Here in fact, Gluck put into practice a dramaturgical principle that he had announced ten years before in his preface to *Alceste*; and it was never again to be so successfully carried out as here until Wagner's day. The music of both the storm and the Furies' apparition belongs of course to the Rameau tradition, employed with less musical art than Rameau's but far greater dramatic forcefulness. The neat contrasting of the Scythian ballet and chorus, raucously violent and barbarian, with the serene euphony and shapeliness of the Greek priestesses' music, is also French—though Gluck had made a similarly telling contrast between Spartans and Phrygians in his *Paride ed Elena* of 1769. The heroic, orchestra-dominated rhetoric of Pylades's 'Divinités des grandes âmes' in Act III may even seem to some listeners to introduce an unwelcome note of (again French) theatrical inflation into the economy of the drama.

The scene in which Iphigenia and Orestes eventually recognize each other, as she is in the very act of immolating him on the goddess's altar, is a triumph of musical and dramatic compression. Euripides had engineered this recognition by means of a letter communicated to Orestes for no very convincing reason; and Goethe achieved sublime psychological truth by making the confession spontaneous. Guillard combined psychological and dramatic truth when he made Orestes murmur, as the sacrificial knife is about to descend: 'It was thus that you perished in Aulis, sister Iphigenia'—something so touching and so completely natural that it seems a self-evident solution to the problem. Here, as in the Furies scene, Gluck borrowed music from an earlier work. The eight bars in the orchestra, a shifting chromatic A minor with strong Neapolitan colouring, which accompany Iphigenia's approach to the altar, are taken from his ballet *Semiramide*. A pre-romantic string tremolando depicts her shuddering grasp of the knife; the first violins' sforzando notes and the Neapolitan B flat which forces A minor into D minor express her anguish and are echoed in the calm words of Orestes, which modulate back to A minor—it is all over in twenty-five bars of music.

In all performances of this, and any other of Gluck's operas, the recitative is the chief stumbling-block. It is very difficult for a singer today to invest Gluck's stately and, as it seems to us, unnaturally slow-moving dialogue with the right weight and dignity without seeming pompous. The quick, conversational secco recitative of Mozart's comedies is easy and natural by comparison. For Gluck's recitatives French singers have the great advantage not only of possessing the language itself but also of belonging to the same dramatic tradition, the tradition of Racine and Corneille on which Gluck consciously modelled his style. English audiences who find Racine dull and unnatural can hardly fail to find Gluck still duller and more unnatural. Even Gluck's use of modulation, to point or emphasize a change of mood, may well pass unnoticed except by those who are able to listen, as it were, historically. So that, for example, we easily lose today the deliberate rhetorical effect of the stately ten-bar sequence in which Pylades, at the opening of Act II, moves from G major through E major to A minor; and the contrast between this simple diatonic progress and Orestes's swift interruption with the first of a series of diminished harmonies or dominant sevenths. For Gluck's contemporaries the passage expressed clearly the sanity and nobility of the one and the tense, neurotic over-excitement of the other. There is no way round this problem of recitative, whose solution demands the same qualifications as in Gluck's day—a full command of classical French declamation, a beautiful and above all a noble quality of voice, and a sense of the importance to be attached to every shade of dynamics and expression. No wonder that it is much rarer to hear the recitatives of Gluck's operas convincingly sung than the formal numbers.

The part of the chorus—Greek priestesses, Scythian warriors and crowd, and the Furies—is much more than decorative or incidental. The characterization is chiefly rhythmical. The Scythians sing and dance in very simple, even crude duple time, but the longest of the priestesses' choruses, at the end of Act II, is a graceful 3/8; and it is only their slow processional, almost liturgical music that is in duple time. The music of the Furies—D minor in common time—is distinguished from that of the Scythian warriors by its fierce harmonic progressions and the relentless upward scale of the trombones cutting across the voices. It was direct, unambiguous, forceful passages such as this that made Gluck's great reputation in his own day. After *Orfeo* and *Alceste* he was consciously the architect of the new music drama, just as consciously in revolt against opera conceived as a series of vocal displays as Wagner was to be eighty years later. The movement to reinstate the drama, to give that drama unity and close continuity and to prune it of extraneous diversions, delightful in themselves but strictly irrelevant—all this had begun at least a quarter of a century earlier. Although it originated in Italy, the champions of this movement had always looked to French opera, where the singers had never made comparable claims, as a model. It was in the French court at Parma that the first 'reform' operas were given, and only later in Vienna. But it was in the cosmopolitan atmosphere of Vienna, though still with an Italian for his librettist, that Gluck first brought to fruition these earlier sporadic attempts to reinstate the drama. He was as clear, and almost as articulate, as Stravinsky in explaining and justifying the aims of his music; and this clarity and articulateness served him particularly well when he encountered the French journalists and littérateurs.

As often happens in history, the quality which made Gluck's reputation in his lifetime has proved to be the quality which now most clearly dates his work. The simplicity and directness of Gluck's musical language and the absence of all specifically musical interest—any inorganic musical elaboration, that is to say, anything not strictly necessary to the drama—can easily sound heavy and monotonous today. Compared with the musical variety and resourcefulness of Rameau, Gluck is musically a dull dog: it is as a music-dramatist that he is so immeasurably superior. In Iphigenia's air at the end of Act II, 'O malheureuse Iphigénie', the shapes and the texture of the music remain totally unaltered for over a hundred bars, with four crotchets to the bar—the third of which is always accented—forming the bass line. Modulation shows equally little variety. It takes Gluck twenty-five bars to reach the dominant and another twenty-five to return, after no adventures whatever, to the tonic. The remaining fifty bars touch the subdominant and the tonic minor, but venture no further afield. To the modern ear these present large stretches of virtually undiversified sonority—in fact they can easily sound literally monotonous. Gluck's justification, I feel sure, would be twofold: first that during this air Iphigenia's state of mind never changes—she pours out her grief without expatiating on it; secondly, that it is for the singer concerned to infuse this single idea, in its virtually single formulation, with a sensuous beauty and expressiveness of tone, word and phrasing that should make all decoration, all elaboration either vocal or orchestral, unnecessary.

And now one word about the final scene, in which the goddess Diana appears—a *dea ex machina*—to reverse the obvious conclusion of the drama. Obvious in the sense that even if Pylades succeeds in killing Thoas, it is unthinkable that a handful of Greeks could have fought their way through Thoas's assembled guards and a mob of bloodthirsty barbarians. The Greek contingent may exclaim, in Guillard's words, 'Of this odious tribe let us exterminate all down to the last remnant!', but their hope of doing so would have been small. And so it remains for the gods themselves to rectify the balance and rescue Iphigenia and her brother. If this seems to us now tame and contrived—and very unlike our own experience of life—it should not be forgotten that the whole framework of the story is supernatural. Iphigenia herself had been mysteriously wafted away from the altar when her father was about to sacrifice her in order to obtain favourable winds for the Greek fleet sailing to Troy. She was thus in every sense Diana's protégée as well as her priestess; and it was Diana's brother, Apollo, who had despatched Orestes to bring *his* sister back to Greece and Diana's statues from Tauris to Athens. The brother-sister theme in fact shapes the drama on the supernatural as well as the natural plane; and so Diana's appearance is much the less arbitrary. In any case *Iphigénie en Tauride* is a work that can only be approached by those who are prepared to accept what is in effect a triple, often overlapping range of conventions—those of Greek tragedy, of French classical drama, and of eighteenth-century opera. For them this is one of the greatest prizes and immeasurably worth the labour of some mental adjustment.

Orfeo[*]

Let us first get the facts rights—and they are strange enough in all conscience. Gluck wrote his *Orfeo ed Euridice* for the celebrations of the name-day of Maria Theresa's husband, Francis of Lorraine; and this circumstance conditioned both its length and its character—the work must not be too long and it must have a happy ending. The presence of an original-minded ballet-master at the Court Opera, in the person of Gasparo Angiolini, and the unusual character of the librettist, the literary amateur and subsequent adventurer Raineri Calzabigi, were even more influential in deciding the nature of the work. And it

[*] BBC 3 June 1972.

was therefore a concatenation of chances rather than deliberate policy that gave *Orfeo* its final shape—a short *festa teatrale* of courtly character, with many ballet *divertissements* and an important part for the chorus in the French style. The fact that the overture is hardly more than an orchestral summons to settle the audience in their places and that the last act is musically inferior to the other two, terminating in the not particularly interesting celebrations, makes the actual body of the work still smaller. For the part of Orpheus Gluck engaged the castrato Gaetano Guadagni, a quite unusually intelligent singer who had worked for Handel in England a dozen years earlier, a remarkably handsome man with a fine contralto voice. This he used to obtain effects of nobility and grandeur rather than to display a technical agility which he seems, in fact, not to have possessed.

Orfeo, then, was a work of transition, a halfway house between the twenty or more conventional Italian operas of Gluck's youth and the so-called 'reform' operas which began five years later with *Alceste*. So much is plain; and we should perhaps realize that in reviving the first Italian version of *Orfeo* modified only by the substitution of a woman contralto for a castrato, we are in fact taking into the modern operatic repertory a unique example of an eighteenth-century court entertainment. The exact importance of substituting a woman for a castrato I shall consider in a moment. Twelve years after the first performance in Vienna, the Paris Opéra was certainly very clear that some substitution was absolutely essential if the work was to be performed there. For France was the one country in Europe that had never accepted mutilated singers— presumably for the reason voiced by the Président des Brosses that 'it is not worth forfeiting one's effects for the right to chirp like that'—and also perhaps from a sound, fundamentally democratic *bon sens*, an instinctive feeling that the castrati were in fact Italian boys who had been, literally, butchered to make a Roman holiday, sacrifices not so much to the Muses as to the refined tastes of a tiny minority. Gluck therefore re-wrote, transposed, and elaborated his original score, arranging Orpheus's arias for a tenor voice and rewriting the recitatives to fit the French text. And so there existed two quite distinct versions of the work until 1859, when the opera was revived for the mezzo-soprano singer Pauline Viardot. Berlioz, who was entrusted with the musical side of the production, could have returned to the first, Vienna version; but instead he rearranged and transposed the Paris version in which, as you remember, Orpheus is a tenor. This third version was subsequently translated back into Italian, and it is this adaptation that, with various modifications, is normally performed today. Pauline Viardot was a mezzo, however, not a contralto; and it therefore happens increasingly that contraltos prefer the original version written for Guadagni. It is certainly more historical to perform one of the two versions that Gluck himself composed, rather than a third made almost a century later. But there are objections to this; and the first and most important concerns the entrusting of Guadagni's part to a contralto woman singer.

I should mention that when Paris wanted to hear *Orfeo*, it never occurred to Gluck to entrust the role to a woman singer. For him, evidently, there was no simple substitute for the castrato voice, and he preferred to alter the whole character of the work, to 'de-sacralize' it, so to speak, and to exchange for a merely bereaved husband the mythical figure embodying the power of music, always finding and always losing, existing only in song.

What did the seventeenth and eighteenth centuries understand by a castrato singer? What were the qualities of their voices? This is what Goethe says on the subject:

> I have reflected on the reasons for these singers pleasing me so much, and I think I have found it. In these performances the concept of imitation and of art is invariably more strongly emphasized, and by their great skill a sort of conscious illusion is produced. And so there is a kind of double delight—in the fact that these artists are not women, but only represent women. These young men have studied the properties of the female sex, its nature and behaviour; they know them thoroughly and reproduce them like artists— they represent not themselves, but a nature absolutely foreign to them.

A castrato singer is a stylized human being, able to sing either male or female roles because neither his voice nor his personality (of which the voice is only an expression) seem to belong to either sex. And it is perhaps significant that in the third act of *Orfeo*, where Orpheus is in fact harassed like any other husband who might inexplicably refuse to look his wife in the face, the level of Gluck's inspiration noticeably falls. It is only when

Euridice has once again died that the power of song—of which
Orpheus is the ideal embodiment—returns to him in full
measure. The grief and bewilderment of his music in Act I and
the heroic pleading with the Furies and the expression of Elysian
ecstasy in Act II are only paralleled in Act III by 'Che faro'—
grief and bewilderment again, but wholly unlike the minor-
mode lamentations of Act I. In this surprisingly simple,
unelegant major-mode melody there is a note of transfigured
grief—as though it were only through Euridice's death that
Orpheus can once again embody music. Bereaved, he is a demi-
god; re-united with Euridice, merely a uxorious husband. Like
the figure on Keats's Grecian urn, it is part of his very essence to
remain forever unsatisfied:

> Bold Lover, never, never canst thou kiss,
> Though winning near the goal—yet, do not grieve;
> She cannot fade, though thou has not thy bliss,
> For ever wilt thou love, and she be fair!

The heroic, transfigured character of 'Che faro' makes it wholly
out of place as the expression of mere despair at a wife's death:
its unmistakably triumphant character can, I believe, only be
explained if we realise that for Gluck and Calzabigi the conjugal
relationship between Orpheus and Euridice was only a parable,
a symbol of the illusion that beauty can be possessed and made
permanent. Orpheus, like the artist, knows that beauty can only
be captured fleetingly, on the wing; and that the moment he has
embodied his vision, it dies for him and the search, which is his
life, begins again. In this spirit the poet Rilke apostrophises
Orpheus in the third of his sonnets:

> Gesang, wie du ihn lehrst, ist nicht Begehr,
> Nicht Werbung um ein endlich noch Erreichtes;
> Gesang ist Dasein.
>
> (Song, as you teach it, is no more Desire,
> No quest for something finally attained.
> Song is existence.)

Gluck himself was well aware that the scene of Euridice's
death was precariously poised, particularly with an audience as
alive as the French to the comic potentialities of any stage
situation (no doubt another facet of the quality that made the

French reject castrati singers). 'Taken a shade too fast, or sung without the exact degree of seriousness and pathos demanded by the situation,' he said, ' "Che faro" becomes a merry-go-round tune.' What, then, were the qualities of voice that Gluck looked for in his castrato Orpheus and felt could be replaced better by a tenor than by a woman contralto?

The distinguishing characteristics of the castrato's voice were clearly described by the Président des Brosses, who went to Italy with a strong prejudice against these singers but was more than half-converted:

> Their timbre is clear and piercing as that of choir boys, only much more powerful. . . . Their voices always have something dry and harsh, quite different from the youthful softness of women; but they are brilliant, light, full of sparkle, very loud and with a very wide range.

This certainly resembles a certain kind of high tenor voice more than anything else in our experience; and it is in every way the opposite of the contralto woman's voice. For if a castrato is in effect a stylized human being of neither sex, or even a stylized man, the contralto is felt by most people—and apparently by all composers—to be a severely departmentalized woman. A glance at the operatic repertory will show that contralto roles fall into two main categories. In the first come the mature, maternal characters—Fricka, Erda, Geneviève, and their comic or sinister counterparts such as Marcellina and Mistress Quickly, Azucena and Ulrica; but a much larger group is constituted by the voluptuous and power-hungry women, a category which includes Dalila, Ortrud, Clytemnestra, Amneris, and Eboli—all women in whom femininity is doubly strong for having been in some way frustrated or perverted.

There is never any doubt of the full femininity of any of these characters and Théophile Gautier—a notoriously unmusical man—seems alone in finding the contralto voice bi-sexual:

> Que tu me plais, ô voix étrange!
> Son double, homme et femme à la fois,
> Contralto, bizarre mélange,
> Hermaphrodite de la voix!

For opera composers, on the other hand, the contralto voice seems to be associated with a femininity at the opposite extreme—both in timbre and in psychological association—to that naturally demanded by the part of Orpheus, either as bereaved husband or as personification of the power of music.

Nineteenth- and twentieth-century composers in search of a voice of indeterminate sex, or in some way dehumanised, have never turned to the contralto woman. Wagner, after hearing some of the last castrati to sing in the Sistine Choir, played with the idea of 'enticing one of them away from Rome to play the part of Klingsor', the eunuch-magician in *Parsifal*. Rimsky-Korsakov chose a very high tenor for the part of the Astrologer in *Le Coq d'Or*; and Britten made his Oberon a counter-tenor— quite a different voice from the low-lying, or contralto, castrato. But the voice best corresponding to that of a castrato, the low-lying counter-tenor voice, has neither the volume nor the stability needed for the part of Gluck's Orpheus, and this is presumably why Gluck didn't choose it for his Paris version. Glinka, Stravinsky, and Schoenberg resorted to polyphony—a single utterance, that is to say, composed of several voices—to represent divinity. There seems, in fact, to be no natural association between the contralto woman's voice and the superhuman or the preternatural; and its sexual associations appear to be overwhelmingly feminine, often in a pejorative sense; so that on both counts it is unsuited to the role of Orpheus, who is both a symbol of the power of the music and a noble, bereaved husband.

Of course, our reactions to *travesti* parts—*Hosenrollen*, as the Germans realistically call them—have changed with our attitudes towards sex. Almost since Gluck's own day it has been unthinkable that a woman's role should be played or sung on the professional stage by a man, except in farce; Goethe's was probably the last generation (even in Italy) that accepted this. In modern times it is only under Japanese influence that Britten has given the role of the Mother in his *Curlew River* to a male singer—thereby, as it were, stylizing and universalizing the concept of Maternity, in very much the same way perhaps as Gluck stylized and universalized his Orpheus by giving the part to a castrato.

We accept the soprano heroes of Handel's operas for the music they sing, without pretending that this is a happy solution or one that leaves the drama unimpaired; and unlike Gluck, Handel left no alternative versions obviating this particular convention of the day. Our grandfathers, unused to the unrestricted vision of women's legs that we enjoy, found Verdi's adolescent Oscar, Mozart's Cherubino, and Strauss's Oktavian pleasantly disturbing. The emergence of unisex has removed this *frisson*, though without relieving the directors of opera-houses of the task of finding not simply the right voice, but the right voice combined with the right figure, for such juvenile roles. Here again, as in the case of Handel's operas, we have a convention that we accept for the sake of the music—an historical fact in Handel's case and one based in human physiology in the other—for no adolescent of either sex has the vocal organs needed to portray adolescence on the operatic stage.

In all these cases we are concerned either with a subordinate role, or a single one among several important roles. *Orfeo*, on the other hand, is virtually a single-part opera and it is only revived for individual singers. My objections to the original version of the work being performed with a woman singing the title role should not be taken as any reflection on the great singers who today, as in the past, have taken up this challenge—only as a reasoned statement of why, as I see it, the challenge is in effect an impossible one: there is, I believe, no satisfactory solution but Gluck's own, which was to rewrite the work for a tenor.

On the other hand, ever since Pauline Viardot's performance of Berlioz's 1859 version, this has proved a superb vehicle for the art of a great singer. The periods during which *Orfeo* is not performed are not those in which Gluck's music goes out of fashion, but simply years during which no great contralto singer has appeared at the right time in the right place. For there is no doubt that to interpret this role convincingly, which means in effect to bear the complete responsibility for the whole work, needs not only an unusual voice but even more an unusual artistic personality.

The Royal Opera House has chosen, I think wisely, to present *Orfeo* in an 'antique' rather than eighteenth-century setting, following what was apparently the practice in the first Viennese productions. This means that the scenes round Euridice's grave, the appearance of Amor in Act I, and the 'infernal' and 'Elysian' scenes in Act II are not rendered painfully incredible by the presence of an Orpheus in an eighteenth-century wig, formal mourning dress, and a large decoration such as the Drottningholm Opera gave us recently at Brighton. There are enough anomalies in *Orfeo* without adding others that can be avoided.

Gluck's music on the other hand is free of all anomalies of any kind—a fact which played a large part in its initial success with a public for whom Rousseau's 'nature' and Winckelmann's 'antique' were beginning to coalesce into a new ideal. In his lifetime Gluck was regarded as an innovator; but his music, like Wagner's a century later, proved to be the consummation of one musical era rather than the opening of another. When such things were fashionable, it used to be said of his music that its beauty was 'moral'—which is perhaps rather like saying, in despair, of a plain woman, 'she has beautiful eyes'. The unexpressed identification in that judgement is between morality and simplicity, or between lack of ornament and sincerity; and it is still possible to find English people whose dislike of baroque architecture or rococo decoration is rooted in the same identification.

When *Orfeo* was revived by Solti in 1969, Charles Mackerras produced a version based on the Italian original, but including material from the French version. Mr Mackerras is now himself conducting a version approached, as he says, from the opposite angle. 'Neither of the two authentic versions is theatrically viable today,' he believes, 'so I have based the present revival on Berlioz's version in an Italian translation, on the grounds that it contains all the most famous pieces in their most developed (and most familiar) form and is more suitable for Miss Verrett's voice.' This is wholly reasonable; and I only hope that one day Mr Mackerras will approach the problem from yet another angle and give us a modern 'performing' edition of the French version with the part of Orpheus sung by Nicolai Gedda or one of his successors.

18

SHORT THOUGHTS

Understanding music[*]

'Understanding' music is an ambiguous expression. It is quite possible to study acoustics and the structure of the human ear and brain—the physical mechanisms, that is to say; to master the mathematics of the diatonic, chromatic, and any other scale; the history of musical instruments, forms, and cultures—all this; and yet to remain absolutely ignorant of the essential character of the art, to be—as we say—'unmusical', deaf to the significance of the sound-patterns which the ear reports to the brain.

In what does this significance consist if it can elude the most intelligent and the most learned, yet fill the dullest and most ignorant with enthusiasm? If we compare music to a language, we have to admit that it is unique in not possessing a vocabulary; and in fact that its warmest devotees fail to reach agreement on any meaning that can be formulated in words, and only agree that any such hypothetical meaning would be entirely dependent on context. An odd language, in fact!

If music is not so much a language as a system of patterns comparable to architecture, mosaic, or the innumerable structures of the physical world, we are still left with two major facts unexplained. I mean the possibility of constructing a musical argument, or dialogue, outside the terms of conceptual thinking; and of arousing in the listener a succession of emotions, very strong and very distinct yet only susceptible of verbal expression by the use of crude and vague metaphor.

It is clear, then, that it is its double origin, intellectual and emotional, that presents the chief problem in 'understanding' music; and the situation is further complicated by the fact that

[*] BBC 6 January 1977.

each listener has, as it were, his own receiving-set—so that he hears something different not only from what his neighbour hears but from what the composer had in his mind at the time of composition.

That the matter is not quite so hopelessly subjective as this might suggest is proved by the high degree of consent among musicians and music-lovers. They may not hear exactly the same G Minor Quintet, and they certainly all hear a different quintet from that conceived nearly two hundred years ago by Mozart; yet they all agree in principle that the work is a great one and are sufficiently at one in their views of it either to give a performance or to judge such a performance 'good' or 'bad' i.e. true or untrue to the composer's ascertainable intentions.

Further light may perhaps be thrown on the subject by a negative approach—by considering, that is to say, what we mean when we speak of not understanding a work. Anyone sufficiently interested can of course learn how a work is constructed. What he needs to understand is not this so much as the mentality of the composer: so that what we mean by understanding in this context is really identification, the willingness and ability to identify ourselves in imagination with the author. In differing degrees we have learned to identify ourselves with the great composers of the past and can, at least in some sense, 'sympathize' or 'feel with' Mozart, Schubert, Beethoven, or whoever it may be. But by the middle of the last century, the personality of the individual composer had begun to play a much more important part in determining the character of his music; and controversial figures appeared who divided, and in some cases still divide, the musical world. There is nothing difficult to understand intellectually in the music of Liszt, Berlioz, or Mahler, for example; but these men's personalities, which find vigorous expression in their music, were strong enough to repel as well as to attract, and left very few music-lovers neutral. In a later generation it was not at bottom the new language of Bartók or Schoenberg that made their music unpopular, but their strong personalities (and similarly, the equally strong 'anti-personalities' of Stravinsky and Hindemith, it should be added). All these men inhabited emotional worlds that were felt as alien by their first listeners, who resented what seemed—and often were—flat contradictions of the moral and aesthetic ideas which they mistook for self-evident truths.

It is no different today. If I say that I do not understand Stockhausen's music, what I am really saying is that I have no sympathy with the mental, emotional world from which that music springs. Such a lack of initial sympathy, which makes it impossible for me to identify myself in imagination with the composer, lies at the root of all failure to understand music.

Some composers seem to invite this act of sympathetic identification—Britten is an obvious instance among twentieth-century composers, and Janáček another—while others, like Varèse, seem instinctively to ward off such a relationship. Still others, though they are few, give the impression of remoteness, of an indifference to the potential listener: Webern seems to me the clearest instance of this. Again, such 'elective affinities' are highly subjective and depend on many factors in the individual listener; yet it remains true that without this imaginative link between composer and audience no music can be said to be understood.

Shape and sound[*]

Every human being springs from the mixture of two independent stocks, a father and a mother, and different members of the same family will show a stronger allegiance to one or other of these inheritances. In much the same way music springs from two different forms of the creative instinct, the architectural and the poetic: the creation of shapes and patterns, that is to say, and the reflecting and expressing of thoughts or feelings in sounds. But whereas poetry in the accepted sense involves the use of words, and therefore at least a minimum of conceptual thought, music employs sounds which correspond to nothing in the world of conceptual thinking proper. Their significance is infinitely various and depends on a great many circumstances—rhythm, pitch, dynamics, timbre, and so on—but also, and most importantly, on context.

I remember as a child, whom an obsession with the piano and a perilous facility for sight-reading inevitably labelled 'musical', the shame I felt when intelligent friends of my parents took it for

[*] BBC 30 November 1976.

granted that I must therefore also be a good mathematician. In fact I was, and am to this day, a cretin with figures and have no facility whatever in the theory of music, in which mathematics plays a certain part. Rhythm, pitch, intervals, and indeed patterns are all subject to mathematical laws, and we should never forget that for at least the eight hundred years separating St Augustine from Plato, 'music' was considered a department of mathematics and mathematics a department of philosophy. Unfortunately we have been left only too well informed about the Greek theory of music and almost totally ignorant of the sound of their music or indeed their musical instruments. European art-music continued to be closely related to learned theory—professed for the most part by clerics and concerned almost exclusively with ecclesiastical works—until the Renaissance. There was always, of course, popular music and music for the dance: indeed the troubadours introduced a fresh, basically popular melodic note even into art-music. But it was not until the Renaissance that music ceased to be 'the handmaid of theology', closely wedded to mathematically-based theory; and it only very gradually achieved the expressive power over the whole range of human life it has today. In fact it could be said that until 1600 or thereabouts it was the formal, architectural, mathematically-based element in music that predominated.

With the Renaissance, and man's discovery of himself as a third-world child of nature instead of (or at most as well as) an other-world child of God, all the arts—including music—took on a new human dimension; and human life itself, in all its many-coloured variety, assumed an entirely new importance. The drama and the pathos of human relationships and the awareness of the physical world demanded a new form of the musical language; and the instinctive answers to this demand were the development of our diatonic, or tonal, system in place of the old modes inherited by the Christian church from the Greeks; and the gradual replacement of the essentially communal, many-voiced polyphony by a music in which a single top-line of melody—the voice, as it were, of the newly enfranchised individual—assumed an entirely new importance. The function of the other voices, or lines, was subordinate—to provide an accompaniment, or harmony, which enriched and gave perpetually differing character to the melody. Music now took on a new poetic and dramatic character; and the other, maternal side of its family origins came to the fore, most notably in the new form of the opera and its first great master, Monteverdi. The older, paternal tradition, on the other hand, was never lost. Monteverdi himself used it in his works for the church; and for another two hundred years, at least until 1800, this style, by now consciously archaic, was used for all large-scale choral compositions. It was used by both Handel and J. S. Bach in their instrumental and choral music, though this earned Bach the reputation of an old fogey with his smart, musically up-to-date composer sons, Carl Philip Emmanuel and Johann Christian.

The place where these two traditions achieved an ideal balance was, for reasons both historical and geographical, Vienna; and the unparalleled greatness of the composers of the so-called First Viennese School—Haydn, Mozart, and Beethoven—lay precisely in their ability to exploit both traditions to the full. Mozart and Beethoven consciously reinforced their mastery of the older tradition, now in danger of sclerosis, by their study of the great masters of the polyphonic past—Palestrina, J. S. Bach, and Handel. It was the youngest of these Viennese composers, Schubert, who in his solo songs brought to a perfection never surpassed, and only rarely equalled, the solo song in which the opposite principle—a single line of vocal melody accompanied by the piano—is embodied.

Throughout the nineteenth century, the two principles co-existed. In symphonies and chamber music the perfection of the Viennese balance between the two traditions remained the ideal; but in two of the most popular musical fields—those of opera and piano music—the older contrapuntal ideal was increasingly sacrificed to the pattern of 'a single melodic line with accompaniment'. So that when, some seventy years ago, the time was ripe for the next major development or revolution in music, it took the form of a reaction against both the innovations that the Renaissance had brought—homophony and tonality. That is to say, the contrapuntal, polyphonic principle was re-established at the expense of the single-line-plus-accompaniment; and the diatonic system, which at the Renaissance had replaced the church modes, was first expanded out of recognition and then, in the works of the Second Viennese School, ousted and replaced

by another method of ordering the twelve notes of the chromatic
scale.

During the last fifty years, music has passed through a stormy
period of quickly succeeding, often fundamental changes; and
no point of rest, no generally recognized goal has yet been
reached. Even so, the two basic affinities—with architecture on
the one hand and with poetry on the other—remain; and if there
is to be again a single musical language in Western Europe, it
must be based on some satisfactory balance between the two.

Words into music[*]

Milton called them 'harmonious sisters' and implied the most
august origins in the 'music of the spheres', that cosmic
harmony which survived in his imagination despite his experience of seventeenth-century political life. Like many sisters,
however, 'voice and verse' have been troubled in their relationship by the kind of jealousies that may spring up in any family
between a pair of handsome and spirited girls, each aware of a
good claim to the first place in their admirers' eyes. Claims of
seniority are here meaningless—though I suppose that primitive
man may have hummed to himself, and almost certainly
communicated with his fellows by inarticulate sounds, before he
achieved articulate speech.

No, the real problem is simply this: whether the use of the
vocal chords for singing rather than speech is in itself so rich and
absorbing a phenomenon that it automatically takes precedence
over any verbal text in the listener's consciousness. And of
course there is no single answer. It will depend in the first place
on three variables—the nature of both the singer's and the
listener's sensibility, primarily musical or primarily literary; and
then on the nature of the 'song' itself. Throughout the whole
medieval period music was, at least in theory, the ancillary to
the liturgical text. In practice, however, this ceased to be true as
soon as polyphony developed, in which the multiplicity of voice-
lines effectively obscured the text for the listener. Even the
Reformers' insistence on the absolute primacy of an intelligible
text could not for long repress music's natural tendency to
blossom on its own account, to send out first tendrils and then a

[*] BBC 24 November 1977.

whole foliage of sound, delightful for its own sake and, like any vine, quickly covering the trellis-work provided by the text.

This spontaneously burgeoning nature of music, to which we owe the development of the whole instrumental repertory, is generally reflected in the words which composers have chosen to set. Although there are many notable exceptions, the rule is that the greatest poems—in which, by definition, there is a perfect identity between idea and expression—do not make the greatest songs. A great poem has its own sonorous values, its own music to which nothing can be added and from which nothing should be subtracted. Setting such a poem to music is only marginally less damaging than translating it into another language. In fact a musical setting is, in a sense, a form of translation; so that those who, for example, know Goethe's lyrics only from their settings by Schubert or Wolf, know them only in 'translation'. In the cases of those two composers that translation may well be another work of art comparable in quality to the poem itself. But by far the greater part of all the best song-writers' settings have been of poems not in themselves supremely great or conceived in an intricate pattern of assonances, rhymes, images, or conceptual links that demand appreciation in their own right.

The typical song-text evokes a mood or presents a picture, and its verbal felicities consist in individual phrases rather than in sustained imagery. Such poems suggest spontaneous musical imagery to a composer, who can often enrich and intensify a comparatively bald text. Perhaps the most obvious instances are Müller's verses set by Schubert in *Die schöne Müllerin* and *Die Winterreise*, or Heyse's and Geibel's Italian and Spanish Songbooks set by Hugo Wolf. In all these the poet's vignettes—thumbnail sketches or small genre pictures—are given a heightened intensity and a richness of emotional content often hardly even implicit in the text.

In the twentieth century many composers have come to feel an inherent contradiction in 'setting a poem to music' and have tried to circumvent this. Writing in 1912, Schoenberg suggested that 'the outward correspondence between music and text—as exhibited in declamation, tempo, and dynamics—has but little to do with the inward correspondence and belongs to the same stage of primitive imitation of nature as the copying of a painter's model.' He had himself, he believed, 'completely understood Schubert's songs, together with their poems, from the music alone ... When one hears a verse of a poem, a measure of a composition, one is in a position to comprehend the whole.' And it was on this principle that he wrote his George settings, *Das Buch der hängenden Gärten*. This supra-rational claim to divine, as it were, the whole oak from a single acorn remained a singularity of Schoenberg's but other composers revealed the changing attitude to words in different ways: Stravinsky, for instance, by having the text of his *Oedipus Rex* translated into Church Latin and Britten by setting poems in languages that were largely or even totally unknown to him— French, German, and Russian. Berlioz had already experimented with inventing a nonsense language for the devils in *La Damnation de Faust*, and contemporary composers' breaking up of words into meaningless syllables reflects the wish to emphasize the purely sonorous (that is to say, musical) character of words at the expense of their meaning. This, of course, was only to follow in the path of the Symbolist poets: Mallarmé himself spoke of poetry 'reclaiming its own property from music'.

The opera-house has naturally been the main field in which the rival claims of words and music have been disputed, Richard Strauss even making this the subject of a complete opera, *Capriccio*. In the opera-house, too, many twentieth-century composers have shown themselves similarly aware of the contradiction inherent in setting words to another, non-verbal music; and they have often preferred almost anything to traditional cantabile. Speech, speech-song, melodrama, wordless choruses, and instrumental interludes play important roles in most operas of the last fifty years, beginning with *Wozzeck*, *Lulu*, and *Moses und Aron* and reduced to something like absurdity in Penderecki's *The Devils* and Ginastera's *Bomarzo*, which are music-theatre rather than opera. In England, on the other hand, both Britten and Tippett have continued the tradition of composing straightforward cantabile music for their operatic texts; and Hans Werner Henze, in *The Bassarids*, seemed for a moment even inclined to return to the grand opera formula as developed by Richard Strauss in his last operas.

The relationship between words and music is in fact protean— perpetually shifting in emphasis and character in a way that confirms the suspicion of a deep-lying incompatibility between the two, certainly not felt by the average music-lover but revealing itself increasingly to intellectual analysis and to sensibilities formed in the light of such discoveries.

Fast and slow*

How fast is 'hurried' or 'hard-driven'? And when does 'slow' become 'dragging' or 'funereal'? The answers depend on a variety of factors: objective (such as the size of the hall and the 'gait' of the music, heavy and rigid, or light and flexible) and subjective—differing, that is, from person to person and, we should no doubt add, from age to age. And it is these subjective factors that I want for a moment to discuss.

Our experience of music is part of our experience of life, our physical and mental existence, and it stands to reason that the enormous acceleration of our transport and communication systems during the last hundred (and especially the last fifty) years has been reflected in the rapidity of our mental processes. Our threshold of 'slow-wittedness' has dropped so low, the faculties of quite ordinarily gifted people have been so sharpened by the demands of everyday life, that to our great-grandfathers we should no doubt appear almost preternaturally 'bright'. And 'bright', of course, is just what we have become, partly out of instinctive self-protection but also under the spoken or unspoken pressures of 'get a move on', 'make it snappy', or 'has the penny dropped at last?'.

Ten minutes in an absolutely silent room with no distractions would seem to many people an eternity, and an experience that would very soon become frightening. The large majority need the reassurance of background noise, preferably music with a regular metric beat, an easily predictable shape, and a brisk gait—whether it is a baroque concerto, a pop group or 'Music while you work'. This music is passively heard, not actively listened to; and the habit of passive hearing, if indiscriminately indulged whether by force of circumstance or by choice, gradually erodes the faculty of concentrated listening. (Bird-watching, by the way, which includes listening, may well be part of the reaction against this rapidly acquired ability to discount,

* BBC 6 August 1975.

rather than to concentrate the attention on, an aural experience.)

Classical Western music until perhaps 1950 was composed against the background of what we should consider silence (it would be interesting to know the number of decibels created by the blacksmith whose hammering enraged Wagner when he was working on *Tristan*); and until the second half of the nineteenth century the overall tempo of life, except in a handful of large towns, had changed very little since the Middle Ages. When we perform the music of the past today, we unwittingly transpose it from one time-scale to another. In the 'Todesverkündung' scene in *Die Walküre*, for instance, where Siegmund is told by Brünnhilde that he is to die, the two exchange a stately, symmetrical dialogue in which the same basic material is repeated three times, with variations: today this seems like a slow-motion film, and to many people unbearably circumstantial. Indeed, it was in reaction against this feature of Wagner's music that the Parisian composers of the 1920s—who shared their Italian Futurist colleagues' admiration for speed as such— devised the so-called *opéra-minute*.

If very few of today's younger conductors choose—or perhaps dare—to adopt what seem the very slow tempos inherited from the older generation of conductors such as Furtwängler and Knappertsbusch, I suspect that the compulsion is neither aesthetic nor reasoned, but purely physical: they simply cannot feel a pulse that beats so slowly as a live pulse. It is significant that, for me at least, the greatest of today's Bruckner conductors is Eugen Jochum, a man in his early seventies to whom Bruckner's time-scale still seems to come naturally.

What, may we suppose, were the fundamental experiences determining that particular time-scale? Well, I imagine that very roughly there were three: the gentle but steady pace of the countryman walking through the Austrian countryside; the solemn, slow-paced unfolding of the sacral ballet of the Catholic Mass, as Bruckner knew it, where dramatic climaxes are interspersed with stretches of essentially static, or rather tireless, contemplation; and the peasants' rough, vigorous dancing in the village inn. Of these three life-experiences how many remain as even a memory now, except among the oldest generation? None.

It is the true andante, crucial for Schubert in particular but a central norm in all instrumental music before 1900, that is in the greatest danger of disappearing—and for a very good reason. Although we scuttle from one point to another in towns, and although the young may hike and the old exercise their livers, walking as the natural means of getting from one place in the country to another is all but obsolete. And in the concert-hall the true andante is a correspondingly rare experience: the tempo is either too fast and the gait hectic, or too slow and the gait painfully deliberate.

A comparable change has overtaken slow movements—the old lento, largo, or adagio. From Haydn onwards these were essentially song-movements, often indeed marked cantabile but increasingly assuming the character of poetic meditation or metaphysical contemplation—the original 'song' transmuted and interiorized to become a *chant intérieur*, a self-communing or an outpouring of lyrical feeling demanding what the Germans have called *Innigkeit* or interiority. These movements are still disciplined by the classical symphonist's architectural sense: the syntax is still firm and not yet dissolved into the interjections, sighs, groans, and whispers of the later romantics.

The ideal tempo for such a movement depends on its density or 'specific gravity'—or perhaps it would be truer to say on the specific gravity of the performer. Not only must the composer possess in his inner world 'something to be slow about', his interpreter too must have at least sufficient contact with that inner world—even if he has no direct experience of his own—to recognize and to convey its quality. Jet-set executants and young performers whose natural gifts enable them to command audiences all over the world long before they reach maturity as artists, quite simply lack the qualifications—the *tenue intérieure* or *innere Haltung*—needed to give such movements their full character and significance. Often they exaggerate the slowness of the tempo and the composer's low dynamic marks and so substitute for a classical slow movement a dreamlike, hallucinatory rhapsody in the style of Debussy's *Prélude à l'après-midi d'un faune*.

This losing touch with the implicit character of the music of the past is not confined to matters of tempo. As society and its attitudes change, the very memory of the conventions expressing those attitudes fades—the gestures, the mien, the gait, and the 'tone of voice'. It is very rare now, for instance, to find a singer whose total performance, musical and dramatic, suggests the royal personages of, say, the Verdi/Schiller *Don Carlos*; and on a more trivial plane the recent London production of an eighteenth-century opera showed the master of the house shaking hands with his servants before retiring for the night—a total misapprehension of relationships. In the same way, the note of supplication that marks, for instance, the 'Kyrie' of Beethoven's *Missa Solemnis* and the self-abasing adoration of the 'Sanctus' now demand in many performers a complex process of 'suspending disbelief' and simulating fervour, hitherto only necessary in the theatre.

Can in fact an artist communicate convincingly feelings and attitudes that he has not only never himself experienced, but that have no place in his picture of the world, and to which he has, as it were, neither the time (nor perhaps the will) to discover a means of access?

19

INNER THOUGHTS

The Acquiescence of Jesus[*]

Many people must have wondered in a vague sort of way how the sufferings and death of Jesus came to be known as 'the Passion'. At any rate I'm sure it puzzles many more people than actually accept the Christian explanation of the story. For it's a remarkable and very cheering fact that although the majority of the world rejects the claims that Jesus made on His own behalf—and those claims are indeed absolutely staggering—the account of this week's happenings still arouses a unique interest. To be deeply moved by the death of Socrates may demand a certain degree of cultivation; but no one is too simple—and not many people are too sophisticated—to be deeply moved, and somehow awestruck, by the death of Jesus.

Why should this be so, I wonder? Why should the gospel narrative of the last week in the life of Jesus, from His triumphant entry into Jerusalem as the crowd's hero to His ignominious death on the cross a few days later, possess this personal compelling appeal—as though it somehow held for each one of us, whoever we are, the secret of how life should be lived? And by that I don't mean anything high-falutin' or mystical—but simply and realistically how to make the most of life as it actually is on this earth; what attitude to adopt in our dealings with our fellow-men and to each day's petty problems and events in order to experience life at its richest. To get the most out of life, as we say. For, after all, that is what all of us want.

Of course there are many lessons that can be drawn from the story of the Passion; but I want to concentrate on one single aspect because although it in fact accounts for the name 'Passion', it is not very often talked about.

[*] BBC 29 March 1972.

Passio (passion) is simply the Latin word for suffering; but it also carries the same meaning that we find in the word 'passive'—I mean 'passive' as opposed to 'active'. And if we look carefully at the story, we shall find that during these last days of His life the behaviour of Jesus was indeed less and less initiated by Himself and increasingly prompted by other people's actions. He was obedient to, and co-operated with, events outside His control.

Let's look at the story. Early in the week came the single occasion on which He is recorded as having used violence—when He drove the money-changers from the Temple. They were the commercializers of religion and so for Him the defilers of the sacred, the 'poisoners of the wells'. After that came the supper at which Mary anointed His feet; and then the Last Supper. On that occasion His actions consisted in washing the feet of His astonished friends, and instituting what they—and His followers ever since—believed to be a unique means of contact with Him in the future. His very last action was to give His betrayer the signal he needed. After Judas had gone, Jesus ceased to initiate anything: action was over, His Passion had begun. And from now onwards Jesus is the centre of a violent drama, but a centre as still as the centre of a great storm. His struggle with Himself in the garden of Gethsemane was partly the instinctive shrinking from death natural in any young creature, partly the human resentment of injustice and bad faith. But the struggle ends with Jesus abandoning His own will. 'As Thy will is,' he prayed, 'not mine.' After that he was the passive recipient of Judas's kiss, even rebuking the follower who tried to defend Him by violence; and there followed His silences before High Priest and Roman Governor, broken only by the tranquil answers most calculated to inflame the crowd against Him; and then the flogging, the crowning with thorns, the dragging of the cross to Calvary, and the crucifixion.

Until almost the very end His thoughts were not for Himself but for others—the women of Jerusalem on the way to Calvary; the soldiers doing their job ('Father, forgive them; they don't know what they're doing'); His fellow sufferer, the Good Thief; His mother. But two of the so-called 'Last Words' are poignantly personal—'I am thirsty' and that heart-rending cry, 'My God, my God, why hast Thou forsaken me?' This surely marked the return, with redoubled force, of what had caused Him such agony in the garden—the despairing sense of total, ultimate dereliction and abandonment not only by His friends, but by God Himself. This meant that He had drained the cup to the very bottom. After that, only the recognition of the end—'It is finished'.

What does it mean, this passiveness, this self-forgetfulness, this tranquil uncomplaining obedience to events? I've often wondered how important an element it is in the Passion. Could it perhaps be this that accounts for the story's compelling power over every kind of person?

If we look back for a moment over Jesus's life, we find that this note of obedience, of 'loving attention' to God's will as manifested in the events of everyday life, sounds like an unchanging pedal, a deep organ note, sometimes louder and sometimes softer but always present and determining the harmony of the whole. Even before His birth His mother's acceptance of the destiny that she couldn't possibly under-stand—'Be it unto me according to Thy word'—voiced the same submission that Jesus was to make in the garden—'As Thy will is, not mine'. And all that we are told of the first thirty years (much the greater part, that is to say) of Jesus's life was that He worked as a carpenter at Nazareth and was 'subject' to Mary and Joseph.

What are we to understand by this 'subjection'? Was it not that spirit of unquestioning obedience, the habit of regarding every circumstance of life, however humble, as the manifestation of God's will for Him at that particular moment of that particular day? For a village carpenter it might mean the breaking of an essential tool in the middle of an urgent commission; the inability to buy wood owing to a client's failure to pay; an unexpected patch of bad weather, or a bout of illness that upset the planned schedule of work. The man who had learned to accept such things as these tranquilly—even gladly—found, when faced by the final crisis, that it was all but second nature to Him to accept apparent defeat, and even death, rather than to rebel and to assert His own will.

This co-operation with God's will, the wholehearted and cheerful acceptance of the frustrations, irritations, petty humili-ations, and the myriad pinpricks of everyday life, is in fact the

secret of that profound, tranquil joy that is one of the chief
hallmarks of sanctity. The French writer Simone Weil has put
this very well: 'Whatever difficulty we have to surmount,
however great our activity may appear to be, there must be
nothing comparable to muscular effort—but only waiting,
attention, silence, immobility, constant through suffering and
through joy. The crucifixion is the model of all acts of
obedience.' Or, as a much earlier French writer put it—'Yes,
Father, yes, and always yes!'

This submission to a will not our own, this saying 'yes' to
every circumstance of our life, may seem to you too negative a
principle, too passive. 'All right for saints and other unpractical
people,' you may say, 'but it won't get you far in the world as it
is today.' Well, the world as it is today is not in this matter any
different from what the world has always been. And the
alternative to accepting and, as it were, using the pinpricks of
everyday life is to waste our energy and lose our peace of mind
by resenting and fighting against them. For nothing will procure
us a life without these petty frustrations and irritations. I'm not
of course speaking of real injustice or other evils which must be
resisted but simply of the trivial circumstances of every day.
These, after all, are the little frictions that wear down people's
vitality and spoil the quality of their living. Jesus's own life and
character make it very clear that His attitude of obedience and
docility can be combined with activity, initiative, and influence.
After all, it was not a harmless dreamer or a rapt contemplative
that the Jewish authorities feared so much and the Roman
authorities consented to crucify. This was a man who didn't
hesitate to alienate the powerful, to denounce members of the
establishment, and to declare the necessity for a complete
change of heart. To provide, in fact, the blueprint for the
greatest of all revolutions, aimed at toppling the greatest of all
tyrants, the self. The message of Jesus's life and death is not then
to withdraw from active life in the world, to stop planning,
serving, even governing—but to do all these things in a new
spirit, the spirit of detachment shown in His own life and death.
Detachment means living in the world and not of the world: the
ability, for instance, to put the whole of ourselves into a project
and, if that project fails, to accept failure with the same
tranquillity, the same welcome even, as we should accept

successes. For if the Passion story shows anything, it is surely that what we call 'success' and 'failure' are strictly irrelevant: all that matters is the spirit in which we accept them. 'Pain is simply the colour of certain events.' (Again I am quoting Simone Weil.) 'When a man who can't read and a man who can look at a sentence in red ink, they both see the same colour: but the colour is only really important to the man who can't read. The man who can is far more interested in the message.' Like joy and suffering, success and failure both have their own message for us, if only we can learn to read it and to forget the colour of the ink in which the message is written.

The swift alternation of an exalted joy and an all but overwhelming grief—and the tranquil steadfastness of Jesus in the face of both—is among the most striking characteristics of the Passion story. Let's go back to the moment when Judas left the Last Supper and Jesus knew instinctively that the last act of the drama had begun.

'Now the Son of Man has achieved his glory,' he cried, 'and in his glory God is exalted.' His attitude to the disciples becomes closer. He was always patient with their inability to understand His meaning. He now responds with a kind of gentle humour to Peter's impetuous profession of eternal fidelity; to Thomas's wish to follow Him, Philip's naïve request—'Show us the Father'—and the other Judas's plea for what we should call an immediate 'publicizing' of His claims. Each of these Jesus quietly deflects, and the deep recurrent note of all that He says is that of joy—joy and peace arising from their love of each other and their loyalty to Him. 'All this I have told you,' He says, 'so that my joy may be yours.'

The joy of a man within a few hours of being deserted by the very friends to whom he is speaking, of being tortured to a protracted death? What bitter irony! And yet this week in which the Passion is, as it were, re-enacted is for Christians 'the Great Week'—the greatest week in the year, the preparation for and the celebration of the immense, undeviating, all-pervading joy of the Resurrection. For, in the words of Kierkegaard, 'properly understood, every man who truthfully desires a close relationship to God, and to live in His sight has only one task—always to be joyful.'

If this joy seems paradoxical, it's because it is indeed founded on the greatest of all paradoxes: the belief that by His Passion and death and rising again Jesus has indeed—as He claimed—overcome the world and taken the sting even from death itself. And the worst of all our failures as Christians, far worse than our obvious human shortcomings, is our failure to radiate this joy which is the strongest argument for the truth of what we say we believe. It's a joy that is not only paradoxical but, in the strictest sense of the word, supernatural. And it springs from a habit of mind not to be fully achieved in a day nor indeed in a year and only by very few in a lifetime. But it is the events of this week that, every year, put new heart into us to continue, reminding us of the deepest truth implicit in the story of the Passion: 'That I spent, I had; that I gave, I have; that I kept, I lost.'

PART VI   POETS AND POETRY

20

MUSIC IN THE GERMAN *NOVELLE**

'Once my father took us to a feast . . . and bade me enjoy the delicious dishes; but I could not, whereupon my father became angry and banished me from his sight. . . . My heart full of infinite love for those who disdained it, I wandered into far-off regions and for long years I felt torn between the greatest grief and the greatest love. . . . And so the news of my mother's death reached me and I hastened home. . . . We followed her body in sorrow and the coffin sank into the ground. From that time on I again remained at home. Then my father took me again to his favourite garden and he asked me whether I liked it, but the garden wholly repelled me, and I dared not say so. Reddening, he asked me again,—did the garden please me? I said "no", trembling, and my father struck me and I fled, turning away a second time and wandering far, with a heart filled with endless love for those who scorned me. For many a year I sang songs, but whenever I tried to sing of love, it turned to pain, and when I tried to sing of pain, it turned to love. Thus were love and pain divided in me.' This allegorical tale, generally known as 'Schubert's Dream' and dated 1822 may or may not have been written by the composer among whose papers it was found—in either case it reflects very clearly the style and the ideas of Novalis, the poet and mystic who died even younger than Schubert in 1801, leaving behind him an extraordinary wealth of poems, essays, philosophical reflections, and two *Novellen*— long short stories, one finished and the other fragmentary—in which he foreshadowed many of the ideas and images that were to obsess the German romantic writers. An enthusiastic Christian, Novalis dreamed of *Totalwissenschaft*, or syncretism of beliefs, that should combine the world of medieval religious thought with the findings of natural science, the pietism of the Moravian

* BBC 25 December 1961.

or Herrenhüter sect (to which he belonged) with the erotic idealism and the belief in the primary creative power of art that are central in his vision of the world. In his unfinished *Novelle*, *The Disciples at Sais*, Novalis interpolates an allegory or fairy story—a kind of Platonic myth, in fact—whose hero is Orpheus, a seer who could penetrate the secret unity behind the multiplicity of phenomena:

> He looked for analogies in all things, conjunctures, correspondences, till he could no longer see anything in isolation. All the perceptions of his senses crowd into great variegated images: he heard, saw, touched and thought at the same moment . . . Now men were stars to him, now stars men, stones were animals, clouds were plants; he played with powers and phenomena, he knew exactly where and how to find this shape and the other, to make them appear; and thus he himself drew sounds and melodies from the strings.

Although Novalis speaks little of music as an individual art, his writings are full of musical metaphors. His great Night Hymns find an echo in the love-duet of *Tristan* and his theory of correspondences between the senses prompted not only Wagner's ideal of the *Gesamtkunstwerk*, but many passages in writers as far apart as E. T. A. Hoffmann and Baudelaire, through whom he was to influence first the Symbolists and, at a greater remove, the Surrealists.

The *Märchen*, or fairy-story allegory, was regarded by Novalis as the highest of literary forms because it was nearest to the dream—the state in which human imagination has freest play and can most easily transcend the limitations of the intellect. Little wonder, then, that music—the least earth-bound and most mysterious of the arts—recurs again and again in the stories, the verbal imagery and the mythology of the German poets and prose-writers of the next generation. Goethe had already created a precedent for the insertion of lyrical poems or songs in a prose narrative; and the figures of Mignon and the Harfenspieler in *Wilhelm Meister* represent the classical ideal of music as a heightened, purified, transfigured means of expression—the fine flower, as it were, on the plant of prose, something different not in kind but in function. With the romantic writers, music is something entirely different—a form of magic, an intoxication of the senses, a secret language of the emotions, and a symbol of imaginative freedom and power. In

Brentano's fairy-stories the songs which frequently interrupt the prose-narrative are spells often rooted in folklore and couched in a kind of hypnotic, semi-nonsense language half-way between nursery-rhyme and *Alice in Wonderland* (the 'dream-novelle' in its purest form):

> Schnarch', Karrasper, scharche!
> Schnarassel schnarcht im Sarge.

In Heine's *Florentinische Nächte* we find this sense of music as a magic power diffused throughout. The stories are told to a girl dying of consumption. She lies on the borderland between sleep and waking and Heine's brisk Voltairean prose moves in and out of her dreams. Woven among his factual, journalistic anecdotes of Bellini and Paganini, Heine conjures up the figure of the devil prompting and guiding Paganini's performance in a kind of 'automatic writing' and the unnamed painter for whom 'musical sounds themselves are an invisible sign-language (*Signatur*) in which colours and shapes can be heard'. In the second story the dog, the dwarf, the drummer, and the dancer performing on an embankment of the Thames conjure up a vision that Stravinsky was to embody in *Petrushka*, and the description of the dancer's mime as 'getanzte Privatgeschichte' looks forward to the aesthetics of Expressionism:

It was a dance that did not attempt to amuse the onlooker by exterior movements, steps or gestures. These seemed rather to be the words of a special language which tried to make a specific communication . . . while the trivial accompaniment of drum and triangle seemed deliberately to mislead me and set me on a false trail.

In fact music to the German romantic writers was before all else a mystery, its power immediate in effect as that of a natural phenomenon. This is nowhere more strikingly expressed than in Kleist's *The Legend of Saint Cecilia*, which has as its subtitle *The Power of Music*. The scene is laid in the Netherlands during the Reformation period. Three brothers collect a band of reforming zealots and go to a convent chapel on the feast of Corpus Christi with the intention of creating an uproar during the celebration of Mass and wrecking the church. The nuns hear of their plan but fail to forestall them. The organist nun is at death's door but surprises them by appearing at the last moment and directing the music—described as 'an old masterpiece of the

Italian school'—which has such an immediate and overwhelming effect on the brothers that they are not only converted on the spot but spend the rest of their lives in the madhouse, praying and singing. Was the sick nun replaced by St Cecilia the patron saint of music? Kleist never says so in so many words, and his description of the office is confused; but the miraculous nature of the whole incident becomes increasingly clear when it is revealed that the organist nun never left her bed on the morning in question and died the same evening.

Magic, mystery, miracle—it is the same even for those writers who—unlike Novalis, Brentano, Heine, and Kleist—had some technical knowledge of music, like E. T. A. Hoffmann, or personal knowledge of a great composer, as Grillparzer had of Beethoven. For all of them the intellectual element in music, and even more the mastery of a musical craft, is not simply indifferent but something hostile to the essential nature of the art. For Hoffmann and for Grillparzer the true musician is a simpleton, an innocent, a kind of 'holy idiot', like Dostoyevsky's Myshkin or Alyosha Karamazov—selfless, chaste, humble, unworldly, hopelessly clumsy, and unpractical not only in everyday affairs but even in the practice of his art.

Der arme Spielmann (1848) is Grillparzer's tale of a rich man's son whose love of music has brought him to pauperdom. He hated the violin when he was made to learn as a boy, but turned to it for consolation when rejected by his family and unhappy, through no fault but innocence, in love. In the streets he plays the simplest popular airs and collects a bare livelihood, but his evenings are given over to *fantasieren*—improvising—which is his dearest delight. The narrator visits him in his attic and finds to his astonishment that this 'improvising' is the most pathetic, primitive, and clumsy form of music-making—the passionate savouring of a single note and the simplest intervals—third, fourth, fifth and sixth. In his playing the Spielmann is concerned only with the purest consonance: 'instead of emphasizing the sense and rhythm of a piece he would dwell on the notes and intervals that delighted his ears, even repeating them with a kind of ecstasy'. And his description of the chord of the diminished seventh suggests a kind of aural addiction, an intoxication with the musical sound in itself, quite divorced from melody, harmony and rhythm: 'the neverfailing bounty and grace of a single note, the musical sound in itself: miraculous satisfying of the thirsting, languishing ear'. For the sound-addict, words are a profanation of music and there is no question of rhythm, harmony, or even the simplest melodic shapes.

Where did Grillparzer find the model for such a figure? Some of the Spielmann's experiences are drawn from the writer's own experience. But is it possible that a friend of Beethoven's (and one whom the composer wanted as a collaborator) possessed so primitive a musical organism as this? No, it seems much more probable that Grillparzer was here embodying a common romantic conception of the musician or was, consciously or no, influenced by another fictional musical character—the 'Johannes Kreisler' of E. T. A. Hoffmann. Kreisler the bohemian Kapellmeister, introverted and psychologically unstable, the mockery of his musical friends, first appears in Hoffmann's novel *Lebensansichten des Katers Murr*, where his adventures at a small German court are 'doubled' and interwoven with the autobiography of a professor's cat which has learned to read and write. The *Kreisleriana*, which prompted a set of piano pieces from Schumann who recognised a kindred spirit in Hoffmann's imaginary musician, are supposed jottings from his diary and personal papers. In them Kreisler makes a mystery of his birth and describes himself as 'a new kind of being', one in whose constitution 'too little of the phlegmatic element has been included'. He burns whatever he composes and his chief delight is improvisation. But, like Grillparzer's street-musician, he has absolutely no natural technical facility and his lack of this, coupled with his hatred of virtuosos and their music, earns him the name of *Musikfeind*, music-hater. Only his aunt understands him and his obsession with the beauty of pure isolated musical sounds. 'She spoke of feeling, while my father spoke of understanding music.' There is an account in the *Kreisleriana* of the meeting of a 'musical-poetical club', at which Kreisler improvises on a piano whose upper strings are broken. To a very simple progression of low-lying chords—A flat major, A flat minor, E major, A minor, F major, B flat major, E flat major, C minor, C major—with detailed dynamic gradations he declaims what one of the impatient club-members very justly describes as 'mad nonsense'—*tolle Schnickschnack*—and yet Hoffmann himself does not merely sympathize, he identifies himself with

Kreisler. Although he was an ardent admirer of Beethoven's music during the composer's lifetime, the characters in his stories, like Grillparzer's street-musician, are chiefly interested in old music. One of them, the Baron von B, claims to be the only surviving representative of the old seventeenth-century Italian school of violin-playing. He speaks with enormous authority, dismissing all contemporary violinists except those who have studied with him. Finally, he offers the narrator of the story a lesson, tells him many interesting and useful hints about violin-playing, and at length offers to demonstrate the production of the perfect violin-tone. The result is absolutely pathetic—'like an old woman humming a song with a quavering voice'—but the Baron himself is entranced, in ecstasy. As his pupil leaves, he finds that he has been paid for coming; and yet once again Hoffmann insists that the Baron, far from being a charlatan, is more deeply a musician than the virtuoso 'with his leaps, trills, fast passages, and decoration'.

This hatred of the virtuoso was of course an exaggerated reaction against the taste of the day which was all for 'brilliant execution'—the kind of thing which we today know at its very best in the emptiest keyboard music of Weber and Mendelssohn and the florid arias of Rossini. Even so, it is difficult to understand why any intelligent writer, however ignorant of the technique of music (and Hoffmann was by no means that) should identify the musical character with a kind of idiocy in all matters of the understanding and performance of music. It is almost as though the romantic writers regarded music not only as a secret language of the emotions, which it was profanation to speak with fluency and grace, but even as a kind of hieratic code of which the very syllables, the single notes, were sacred and should be enough for the devout worshipper. When the discontented club-member complained after Kreisler's improvisation that he would rather have heard 'a nice Allegro of Haydn's', the reader is plainly meant to shiver at such uncomprehending Philistinism, and Kreisler himself emphasizes the abstract nature of music in unambiguously religious terms. 'Our kingdom is not of this world,' he says, 'for where do we find in nature any prototype of our art, as the painter and the sculptor find for theirs', and he goes on to elaborate his favourite theory of 'correspondences' in the vein of Novalis:

It is no mere metaphor or allegory when the musician says that colours, scents, and beams of light appear to him as sounds, and that he experiences their mingling as a miraculous concert . . . the sudden promptings in a musician's mind, the birth of melodies, are the unconscious—or rather the verbally inexpressible—recognition and grasping of the secret music of Nature as the principle of all life and activity.

We are not very far here from the Platonic conception of an abstract 'music beyond music', that harmony of the spheres in which idealists have believed, from Saint Augustine to Bruno Walter in our own day, while a belated follower of Novalis, in his belief in the magical powers of music, might be found in Scriabin.

Those who are impatient with such metaphysical fancies might compare two of the most famous Mozart-stories of the German romantic era—Hoffmann's *Don Juan* and Mörike's *Mozart auf der Reise nach Prag*—and ask themselves which is the more revealing, in fact the truer of the two. Mörike painstakingly and, it must be said, charmingly recreates the exteriors of the scene—the language, the clothes, the social attitudes, even the composer's personal mannerisms, as he travels by coach from Vienna to Prague for the performance of *Don Giovanni*. This is a skilful genre-piece, in which factual accuracy is combined with historical imagination. Hoffmann's story, on the other hand, is a penetrating, however phantasmagorical account of a deep musical experience. The traveller who finds himself attending the performance of *Don Giovanni* by an Italian company in a small German town and receives a mysterious visit from the Donna Anna, is no ordinary music-lover. Even if we divest the story of its literary form and its uncanny atmosphere of mystery, the musical and psychological penetration of Hoffmann's comments were unique in his own day and have hardly been bettered since. No one has ever analysed more convincingly the true character of Don Juan or the true tragedy of Donna Anna. Such pages as these and Grillparzer's go a long way to reconciling even the most hard-headed music-lover to these mostly half-forgotten enthusiasts of another age and another mental world.

21

E. T. A. HOFFMANN*

'The higher life of the spirit,' exclaimed the Princess Hedwiga, 'is rooted in a dichotomy, in the tension between the most diverse emotions and the most incompatible sentiments!' She was speaking not only for herself but for her creator, giving expression to a deep-seated conviction which was Hoffmann's most personal contribution to the romantic canon. Duplicity and mystification, ambiguities and magical anatomies, animated puppets and philosophizing animals—Hoffmann's concern was always with the shadows cast by human existence upon the screen of his imagination rather than with human existence itself. He is the poet of the *Doppelgänger*, who was haunted in his own life by a malicious double and wasted much of his energy in a vain attempt to harmonize his real with his imaginary existence. Most bitter of all was his search for a means of self-expression, the struggle to 'double' his life as a civil servant with the life of a practising musician, conductor, and composer; and, when that struggle was eventually successful, the disillusion with his own musical powers which eventually led him quite late in life to writing.

He was the child of a marriage which came to grief two years after his birth and he was brought up by an uncle whose total lack of imagination only exaggerated the bohemian tendencies which he inherited from his father, a Königsberg lawyer. Many years later he put into the mouth of Johannes Kreisler, the most autobiographical of all the characters that he created, a fantastic story whose classical Freudian imagery reveals something of the atmosphere in which Hoffmann grew up, an atmosphere of inhibition from which he escaped by constructing a dreamworld of his own.

'As a boy,' says Kreisler, 'I was not allowed without my

* BBC 5 June 1951.

father's permission into the wood; but the tree and the stone—a murdered girl was buried beneath this stone—attracted me irresistibly. Whenever the little gate in the garden wall was left open, I used to slip out to my beloved stone and I never tired of gazing at the strange patterns of mosses and plants which covered it. I often imagined that I understood the meaning of these patterns, as though they illustrated the stories told me by my mother; and as I gazed at the stone, I thought involuntarily of the song my father sang almost every day. . . .'

It's not surprising perhaps that even as a young man he was afflicted by forebodings of death and haunted by the idea of a *Doppelgänger*, his imagination shaped by repeated reading—'perhaps thirty times', he says—of Rousseau's *Confessions*, the handbook of romantic self-analysis. In a letter written on his twentieth birthday he mentions Jean Paul, the writer who was to have most effect not only on his style but on his aesthetic ideals; for it was Jean Paul who cultivated what Hoffmann was later to note in his diary as one of his most frequent moods—'romantic and capricious to excess, a state of exaltation bordering on madness'. Madness, too, haunted Hoffmann's imagination, whether asleep or awake; and he deliberately cultivated just that simultaneous experiencing of the most diverse emotions which Schumann so much admired in Jean Paul—a kind of amateur, self-induced schizophrenia which had such fatal results in Schumann's case. During his early love-affair with a married woman at Königsberg Hoffmann wrote to his friend Hippel, 'every emotion that I feel for Cora is immediately muted by some comical joke and my heart-strings are so damped that the sounds they give are inaudible'. Hoffmann's metaphors are already musical, but his career at first was almost ludicrously prosaic. After leaving Königsberg in 1796, he spent two years in the depths of provincial Silesia and a further two in Berlin, before finally qualifying for the Prussian Civil Service. In 1800 he left for seven years' state-service in Prussian Poland—two years at Poznan, two at Plock, and three in Warsaw. Already in Warsaw composing and painting took up more of his time than administration, and after a second short stay in Berlin he obtained his first musical appointment—that of 'theatre conductor' in Bamberg. The diary which he kept during his Bamberg years, between 1808 and 1813, is starred with butterflies, wine-glasses, which bear wings on occasions of special 'exaltation', private abbreviations and odd Polish and Italian phrases—all a system of (remarkably transparent) mystification enjoyed for its own sake but consciously employed to conceal from his wife the nocturnal exploits, both bacchic and erotic, in which he tried to forget the day's disappointments. For he was hysterically in love with a sixteen-year-old pupil, Julia Marc; and he was disappointed in his music work, bringing himself slowly and painfully to face his own lack of originality as a composer.

None of the dichotomies in Hoffmann's life was more striking than that between his creative work as a musician and writer. He was a passionate (*schwärmerisch* is the appropriate word) student and admirer of the musical past—of the Italian seventeenth-century masters, of Gluck, Haydn, and above all Mozart—and one of the earliest and most vocal admirers of the 'new music' of Beethoven. But this very admiration was his most serious handicap when he started to compose. The fantasy, the poetry and the wit, which were to make his name as a writer, never find expression in his music, which is in fact no more than a close and often clumsy imitation of his models. Even *Undine*, his single operatic success and praised by Weber, contains hardly a spark of that romantic imagination which had delighted the public in Fouqué's fairy-tale, but was not to find musical expression until Weber produced his *Der Freischütz* five years later.

Nevertheless it was through music that Hoffmann came to writing. In 1809 his *Ritter Gluck* was accepted by Rochlitz for the Leipzig *Musikalische Zeitung*; and with this and his fantastic *Don Juan* story, Hoffmann satisfied both his impulse of ardent veneration for Gluck and Mozart and his need to express himself in writing. During his last years at Bamberg he wrote *Fantasiestücke in Callots Manier* and *Nachtricht von den neuesten Schicksalen des Hundes Berganza*, a continuation of Cervantes's story.

Only when he visited Dresden and Leipzig in 1813 did Hoffmann's imagination suddenly blossom in unmistakably personal form. Here he found himself the involuntary spectator of a bitterly fought war, daily faced with the reality of 'those hacked and mutilated corpses that have always haunted my dreams'. It was in this atmosphere of violence and uncertainty that he wrote the most fantastic of all his tales, *Der goldne Topf*; and finally in Berlin, where he was to spend the last eight years of his life, the spirit which, like some bottled homunculus, had been seeking in vain for escape in his music, finally burst loose and spread itself in a galaxy of tales, novels, night-pieces, and fairy-stories. By the time his opera *Undine* was performed, he had already transferred his primary allegiance from music to literature, and it was among writers that he made his closest friends—Fouqué, his librettist, and Chamisso, by whose *Peter Schlemihl*—the story of the man who sold his own shadow—he had been so impressed.

Das Majorat and *Der Sandmann*, translated during the 1820s, formed the text of a sermon on the evil of introducing the supernatural into fiction. Altogether Hoffmann's success in England was never great. The reason may have been that to a nation that with *The Castle of Otranto* had virtually invented the medium, Hoffmann offered nothing new. In France, however, according to Gautier, he was more popular than in Germany. That popularity was eventually, of course, to determine Offenbach's choice of Hoffmann's tales as the subject for his one serious opera; but it was also to colour the imagination of many of the French romantic artists. Hugo's Quasimodo, for instance, is a Hoffmannesque figure and, consciously or not, Berlioz's *idée fixe* and indeed the whole programme of the *Symphonie Fantastique* are pure Hoffmann. For composers who found nothing to learn from Hoffmann's music have almost up to the present day divined an element in his literary imagination that demands musical fulfilment. Perhaps it is that Hoffmann was the first, not to feel but to preach in words, music as the art *par excellence* of the 'misfit', of the morbidly emotional man or woman, the neurotic haunted by forebodings of 'death' and 'mutilated corpses', of those who share the feeling which he confided to his diary for 1811: 'Destruction is hanging over my head and I can do nothing to avoid it.' His own Kreisler and the Princess Hedwiga of *Lebensansichten des Katers Murr* are characteristic nineteenth-century musical types, introverted and psychologically unstable, who find it difficult or impossible to come to terms with everyday life and seek in music for an escape-world such as no other art offers. Later in the century such people became ardent Wagnerians, and many German courts could boast of a Kreisler and a Princess Hedwiga, their passions of resentment and frustration satisfied vicariously in *The Ring*, their longings for an impossible love immortalised in *Tristan*, and their religious emotions exaggerated in *Parsifal*.

The first and most ardent Hoffmann-admirer among composers was Schumann, who found in Johannes Kreisler something that he recognized as a soul-portrait of himself, involved in mystery and contradictions, Byronic cynicism alternating with a tremulous emotional idealism *à la Jean Paul*. A whole world away from Schumann but bearing eloquent witness to the hold obtained by Hoffmann's peculiarly Germanic, even Gothic fantasy in neighbouring France and Russia, Delibes and Tchaikovsky based their ballets of *Coppelia* and *The Nutcracker* on his stories. And in our own day Busoni and Hindemith have turned to the same source for their opera-books of *Die Brautwahl* and *Cardillac*. At a further remove, Ravel's *Gaspard de la Nuit*—suggested by the writings of a French follower of Hoffmann, Aloysius Bertrand—can be claimed as belonging to the canon; and certainly 'Scarbo', its third movement, is a wholly Hoffmannesque vision.

By investing the horror-stories of Monk Lewis and Mrs Radcliffe with a wealth of Gothic ornament and fantasy, by deepening and enriching their psychological interest and hinting at philosophical, even metaphysical, implications, Hoffmann added a new province to the imagination. It was a province soon to be claimed from across the Atlantic by Edgar Allan Poe, who exploited much of the wealth which had remained unrealized in Hoffmann's lifetime and was succeeded in his fortune by his French discoverer, Baudelaire. The indebtedness of Poe's *Tales of Mystery and Imagination* to Hoffmann is a commonplace; but Poe learned more from Hoffmann than is to be found in his tales; and much of what Baudelaire actually learned at second-hand from Poe he might well have found at first hand in Hoffmann. Listen, for example, to Kreisler expounding what was to become Baudelaire's famous theory of the 'correspondences' between the arts:

It is no mere image or allegory when the musician says that colours, scents, and light-rays appear to him as musical sounds, that their blending is to him a magnificent concert. An ingenious scientist has explained that hearing is a form of internal sight; and in the same way sight is to the musician a form of internal hearing, a means to the deepest consciousness of the music that streams from every object that presents itself to his eyes, and vibrates to the same rhythms as those of his spirit.

Or in a simpler, more concrete form: 'The scent of dark-red carnations affects me with an extraordinary magical power. I sink involuntarily into a dream-like state and hear in the distance the deep, swelling and fading tones of the basset-horn.' This deliberate confusion of sense impressions is but one more example of Hoffmann's instinctive desire to 'double' every experience, to conceive of existence as on two simultaneous planes of which that which is conventionally considered the less real, the shadow- or imagination-plane, is the more intensely experienced. It is in what is probably his best long work, *Lebensansichten des Katers Murr*, that Hoffmann carries this principle to its furthest extent. This is the art of the gargoyle and the grinning Miserere seat, which form in a Gothic cathedral the complement to the statues of heroic saints, crucified Christ, or Virgin Mother. And, just as in the cathedral we sometimes turn with relief to the all-too-human revelations of the Miserere, there must be many readers of *Lebensansichten des Katers Murr* who tire of the noble and slightly fatuous Prince Irenaeus and his *exaltée* daughter, of the amiable Coppelius-like figure of Master Abraham and even of Kreisler himself, and turn with delight to the adventures of Murr, his friend Muzius, and the poodle Ponto.

It would hardly be an exaggeration to say that all the specifically Hoffmannesque effects have been commercialized. The split personality is a commonplace of the cinema, Kater Murr has suffered the final disgrace of being reduced to Micky Mouse. Olympia and Master Abraham's invisible woman have been adopted by the new science of cybernetics, and the whole shadow-side of life—crime and its detection, insanity, abnormalities of body or spirit—is either treated dispassionately and clinically or exploited by writers for whom a mystery means a murder and imagination consists in the invention of some petty detail of circumstance instead of the elaboration of a new mode of feeling, as it meant for Hoffmann. Hoffmann's reputation is nevertheless ensured, a small reputation perhaps but as unique as the life with which his works are inextricably interwoven.

22

JOSEPH VON EICHENDORFF[*]

The general ignorance of German romanticism in this country is easily explained by the language barrier, for lyric poetry is untranslatable. Its essence evaporates on being transferred, as it were, from one bottle to another, and it was as creators of a new lyric poetry that the German romantic writers excelled. Their only other distinctive literary genre, the *Märchen* or fairy-story, depends so much on the language for its artificial *naïveté*, its preternatural, dream-like quality, and many atmospheric effects that, though formally prose, it hardly survives translation any better.

The only medium in which the German romantic world has in fact become known to the non-German speaking world is that of music; and it would hardly be an exaggeration to say that the German *Lied* is as much a creation of the romantic poets as of the composers. Goethe was still wholly a man of the eighteenth century in his attitude to music. He regarded a song as a lyrical declamation with discreet musical accompaniment; and although there are scenes in the second part of *Faust* which can truly be called operatic, there is no instance in his poetry of the characteristically 'musicable' atmosphere of the true romantics—*naïveté*, mystery, nostalgia clothed in neo-medieval imagery, or in symbols drawn from the mystical or metaphysical writers. Even a member of the older generation of romantics like Novalis remains rooted, as a poet, in the purely eighteenth-century world of Young's *Night Thoughts* and intensely subjective, pietistic religion. But Novalis died before the appearance of the richest of all source-books of German romantic writing—the three volumes of German folk poems and ballads published by Arnim and Brentano with the title of *Des Knaben Wunderhorn*, between 1805 and 1808. That the originals were in many cases

[*] BBC 27 December 1957.

faked or drastically 'restored' mattered not at all as far as their influence on poets went—and without this collection the poetry of Eichendorff and Mörike and much even of Heine is unthinkable.

There is very little medieval, 'folkish', *Wunderhorn* romanticism in the songs of Schubert, who chose his poems from minor and already old-fashioned Viennese poets, from the eighteenth century or the first generation of romantics. Of his six Heine settings, for instance, only 'Der Doppelgänger' belongs unequivocally to the psychologically ambiguous night-world of romanticism. It is not until Schumann's songs of 1840, including most of his settings of Heine and Eichendorff, that we see the full impact of *Wunderhorn* romanticism on the song. From then onward, through Wolf and Mahler and in more obscure quarters almost up to the present day, this vein has persisted as one of the strongest in German song-writing.

Heine used folk-song forms and medieval imagery for his own purpose. They were a mask consciously adopted and sometimes derided, even parodied. But for Eichendorff this was the real world of poetry, in which he moved naturally, without self-consciousness or any hint of Heine's cynical *double entendre*.

His range as a poet may be comparatively small but his sincerity is absolute. Nature for him is friendly and harmonious, even in its mysteriousness. His Christian piety is not a neo-medieval stage property, as it was for Sir Walter Scott and the editors of the *Wunderhorn*, for whom Catholicism was picturesque and 'old world'. Even the aristocratic ideal of the medieval knight was more of a reality to this son of a Silesian baronial family than to the urban and middle-class poets who exploited its historical charm.

Eichendorff's life followed the expected pattern of his age and class. He was born in 1788, attended the university first in Halle and then in Heidelberg, where he was on the edge of the Arnim-Brentano circle. He fought against Napoleon like a good patriot and then, in 1816, received a post in the Prussian civil service, living chiefly in Breslau until his retirement in 1844. His first *Novellen* date from 1812, the best known *Aus dem Leben eines Taugenichts* ('The life of a ne'er-do-well') from 1826, and he published his collected poems in 1837. They were new to the public, in fact, when Schumann set a dozen of them three years later. His favourite subjects represent the greatest possible contrast to the life of a Prussian civil servant; nostalgia for a youth spent in the country and the free, untrammelled bohemian life that he can only have known as a student and perhaps as a soldier. His countrysides are central European, with the forest as the image of mystery, romance, and danger—the role played by the sea in the poetry of maritime countries.

> Mir aber gefällt doch nichts so sehr
> Als das deutsche Waldesrauschen

(But I love nothing so dearly as the whispering of a German forest.) *Waldesrauschen* and *Waldeinsamkeit*—the sound and loneliness of the deep forest—were genuine experiences for Eichendorff, not poetic conventions. Sometimes the loneliness is interrupted by a fanciful medieval scene, as in 'Waldesgespräch'— a kind of Belle Dame sans Merci with the hunting-horns sounding in the distance and brought to life in Schumann's setting. Or 'knights in glittering armour' appear as a vision which suddenly fades, leaving an overwhelming sense of terror as night falls—the true romantic *Schauer* suggested in Schumann's setting of 'Im Walde'.

Night—the antithesis of the 'every day'—recurs again and again in the poetry of Eichendorff, as in that of the other German romantics. 'The far distances of memory, the longings of youth, the dreams of childhood, all the short joys and frustrated hopes of a lifetime, crowd around us after sunset, in grey garments like the evening mist,' wrote Novalis. 'Must the day always dawn again? Will the power of earthly things never end?' Mystery, secrecy, spiritual renewal, the return to childhood innocence, the religion of erotic consummation half veiled in metaphysical, archaeological or even botanical symbols—we find them all in Eichendorff; and this nocturnal element in his poetry was seized on avidly by the composers. It is seen at its fullest and most magnificent in two of Wolf's songs—'Verschwiegene Liebe' and 'Nachtzauber'.

In 'Nachtzauber' all Eichendorff's favourite images are combined and each is suggested with wonderful skill by Wolf's music. We hear the murmur of the forest and the streams which run to feed the solitary woodland lakes, with their marble statues. Night descends, bringing the memory of old melodies heard in a dream. In the moonlight a flower stands—or is it a girl? The nightingale, wounded to death by love, sings of happy days long past. 'Come, come to that silent glade.'

There is something of the same spirit, though infinitely less intense and richly variegated, in Schumann's 'Schöne Fremde' (*Liederkreis* Op. 39 No. 6) with its characteristic couplet:

> Was sprichst du wirr, wie in Träumen,
> Zu mir, phantastische Nacht?

But as a rule Schumann instinctively chooses the more innocent, straightforward night-poems of Eichendorff, as in the amazing miniature of 'Frühlingsnacht' (Op. 39 No. 12). Here the passionate excitement of the poet is communicated at once by the racing repeated chords in the piano accompaniment—a spring night in a garden, the first flower scents and the birds returning from the south, the poet half in tears, half laughing with incredulous ecstasy, as the old miracle renews itself.

In another of Schumann's Eichendorff songs, 'Mondnacht' (Op. 39 No. 5), the beauty of the moonlit night prompts the same kind of religious emotions in the poet that we find in Tennyson's *St Agnes Eve*, with the soul 'spreading her wings to seek her true home in heaven'—once again no mere poetical fancy to the genuinely religious Eichendorff, whose complete sincerity communicates itself to Schumann's music. 'Zwielicht' (Op. 39 No. 10), on the other hand, is an unusually ambiguous, allegorical poem of anxiety. Its four verses are set strophically for the voice, though the piano part varies in complexity from a Bach-like chromatic style to simple accompaniment in the last verse: 'Twilight spreads its wings, the trees shiver; clouds gather like oppressive dreams. What does this uneasy moment mean? Have you a favourite hind? Do not let her graze alone. There are huntsmen in the woods. Have you a friend? Do not trust him at this hour. Those tired today are born again tomorrow. Much may be lost in a single night—take care.'

The inhabitants of Eichendorff's poetical world are as remote from those he met in his life as a Prussian civil servant, as the settings of his nature poems are remote from office life in Breslau, Danzig, or Königsberg. The n'er-do-well hero of *Aus dem Leben eines Taugenichts* is a kind of comic Parsifal, a 'pure fool' who acknowledges no civil responsibilities or domestic ties, leaves home to amuse himself and see the world, and gets involved in a succession of picaresque adventures, including a romantic love for a woman whose identity he mistakes. He is a musician—very much the wandering, fiddling *Musikant* and not at all the studious *Musicus*—and some of Eichendorff's most delightful lyrics are scattered throughout the text (once again one thinks of Tennyson, this time of *The Princess*). 'Der Musikant', for example, which Wolf set, says—'I love a wandering life, taking the luck of the road. Even if I wanted to exert myself, I know that it would never do for me.' If girls take a fancy to him, he warns them that marriage would kill his music. This wandering musical tramp is plainly a descendant of the old Harper in Goethe's *Wilhelm Meister*, but a descendant who has shuffled himself free of the Harper's mysterious load of guilt. The wandering scholar is another of Eichendorff's most touching, instinctive expressions of patriotism—which meant a love of German scenery, of northern forest, river, and mountain with a warmth of feeling which no northerner can help feeling when he leaves the Latin south behind him and travels home over the Alps. So bone German is Eichendorff that the foreigners in his stories are hardly more than pasteboard figures, like the Italian painters in *Aus dem Leben eines Taugenichts*, or thinly disguised Germans like the peasants whom he meets on his Italian wanderings. But he would not be the typical Romantic that he is, did he not attempt to portray one of that race who, a hundred and fifty years ago, had a fascination for artists of every description—I mean the gypsies. He succeeds with his genre-picture of the gypsy-girl waiting for her lover and shooting a cat to make him a fur cap. 'He must be dark, with a Hungarian moustache and a gay heart for a wandering life'—Wolf turns the whole scene into a miniature drama in his most characteristic manner in 'Die Zigeunerin'.

Eichendorff has attracted song-writers almost to the present day by the musicality of his verse, its simple imagery, sincerity and strange combination of wisdom and *naïveté* and frankness. Mendelssohn and Brahms each set a handful of his poems and more recently Pfitzner more than a dozen and a whole cantata. The Swiss composer Othmar Schoeck is alone in having based an opera as well as almost forty songs on Eichendorff texts.

Schoeck's opera is based on *Das Schloss Dürande* but no composer has ever made a picaresque opera out of *Aus dem Leben eines Taugenichts*. (This and Schnitzler's story *Casanovas Heimfahrt* are surely gifts to the musician.) Pfitzner's cantata *Von deutscher Seele* dates in its original version from 1921 and consists of almost exactly the same cycle of poems as Schoeck set later in his *Wandersprüche* for voice, clarinet, horn, percussion, and piano. And these poems, as Pfitzner divined, are part of the very essence of the *deutsche Seele*—the German soul. Their simplicity may often sound affected in translation, their imagery conventional, and their sentiments alternately rarefied and naïve. But by his magic use of the German language Eichendorff silences criticism, and the simplicity and sincerity with which he uses what are mere conventions in lesser poets give his works an almost unique place in German literature. Mörike perhaps stands nearest to him, and the two of them played a double role in the later development of the German song, by which they will be remembered far beyond the borders of the German-speaking countries.

23

EDUARD MÖRIKE[*]

'True to the kindred points of heaven and home'—it was tempting to transfer Wordsworth's own skylark image to the poet himself. The mid-Victorian public loved it; and it must be admitted that the poet had done everything, short of actively conniving, to propagate this image of his middle-aged and elderly self. The twentieth century has treated the poetic idols of the Victorian age much as it has treated Victorian furniture—which has been stripped of its heavy veneer—and Victorian buildings which it has cut down to their 'true' scale. Behind Tennyson the seer we have discovered the neurotic, doubt-racked nature-poet and Wordsworth's formidable Establishment façade is no longer allowed to conceal the nature-mystic and the man of passion. The process has been surgical and some of the surgeons needlessly ruthless; but the layers of convention concealing the true poet were perhaps thicker here than elsewhere. The very existence of a Swinburne or a Rossetti made it impossible for the nervous, hero-seeking public to admit a shadow-side to their accepted idols; and no lay person would have dared to write, of, say, the ageing Tennyson as Isolde Kurz wrote in the 1870s of Mörike: 'I felt that this large head of a Swabian country parson with its somewhat flabby features and the deep-chiselled, sullen lines was only a droll or protective mask, from behind which the delicate head of a Greek youth or a smiling Ariel might at any moment emerge.'

Eduard Mörike was born in 1804 and sprang on both sides from the late eighteenth-century professional classes—doctors and clergymen—in the kingdom of Württemberg. He spent his whole life as a country parson, moving from one Swabian village to another, but within these narrow physical limitations

[*] BBC 25 December 1963.

his existence was both complicated and enriched by a number of, as it were, instinctively chosen handicaps. Chief among these was a form of nervous ill-health or hypochondria revealed in the morbid sensibility of some few of his poems, oddly corrected by the tone of the huge majority. A Lewis Carroll-like strain of infantilism kept him to the end of his life emotionally dependent on a younger sister and found exquisite expression in his instinctive understanding of children and tender domesticity, as well as in nonsense verses and some revealing drawings. But unlike Lewis Carroll, he combined with this a capacity for genuine, if often highly spiritualized, erotic passion which found an outlet in a series of obsessions or engagements, and finally in a marriage which would perhaps have killed any man who had not the neurotic's often characteristic toughness and durability. For at the age of forty-seven this Lutheran pastor married a Catholic wife and set up house with her and his adored sister— assuring both women in a sublimely innocent and tactless letter that he found it difficult in his heart to 'distinguish one from the other'. Though apparently bound to fail and certainly swept with storms of jealousy that reduced the poet to despair, this three-cornered relationship and the two daughters of the marriage brought Mörike an enormous amount of happiness; and it was all the more tragic that at the very end of their lives— after twenty years of marriage—husband and wife felt forced to separate.

Mörike appeared to many of his contemporaries very much like Wordsworth's skylark—'true to the kindred points of heaven and home'—for it was the tender simplicity of his poems inherited from folk-models, and those imbued with piety or domestic affection, that delighted the wider public of mid-nineteenth-century Germany. In fact, however, Mörike's domesticity was very far from complacent sentimentality, as we have seen, and his religious faith, though free from any speck of emotional insincerity, was more an ingredient in his poetical make-up than a fully mature or reasoned attitude to life. To his friend and neighbour David Strauss—author of the controversial *Life of Christ* translated by George Eliot—Mörike spoke of his 'permanent inclination to Christianity'; and to Luise Rau, during their engagement, of the chasm between his religious emotions and their public, objective expression:

The gospel offered all its peace and drew me more and more deeply into that solitude of spirit, where the angel of our childhood meets us again and weeps with us. But what I felt belonged only to me myself or to you—I couldn't find a bridge from it to my sermon, and what had *there* been pure gold, *here* becomes dull lead, as soon as I put pen to paper.

The real trinity of inspiration that lay behind Mörike's poetry was a different one. It was not the triune God of Christianity but rather Nature, Love, and Beauty, the trinity that was acknowledged if not proclaimed by the majority of artists in the nineteenth century, in different degrees and with different emphases. Both the simple and flexible language and the clear, strong imagery and atmosphere of Mörike's poems were bound to attract composers in search of song-texts, but the reaction was oddly long in coming. Schumann, the contemporary composer with whom he might seem to have most in common, in fact chose only four of Mörike's poems, Brahms and Robert Franz even fewer; and it was not until 1888, thirteen years after the poet's death, that Mörike found his greatest musical interpreter. Between February and May of that year Hugo Wolf wrote forty-three of the total fifty-seven settings that he made of Mörike's poems and in so doing introduced his Swabian parson—whose fame was inevitably restricted to German-speaking countries—to a univeral public. That contradictoriness that Mörike shared with Goethe, in addition to his lyrical exuberance and purely linguistic virtuosity, found a strong counterpart in Wolf himself. Mörike's poetry belongs to both the 'sentimental' and the 'naïve' of Schiller's categories and he possessed to an astonishing degree the poet's characteristic chameleon-like ability to identify himself with personalities, situations, and atmospheres of the greatest diversity.

Not even Eichendorff was more unaffectedly at home in the folk-poetry world of *Des Knaben Wunderhorn*; and Mörike's natural sympathy with the small, the simple, the humble, and the unsophisticated—that quality that so endeared him to his country parishioners—enabled him to write lyrics which could themselves be mistaken for folk poems. Wolf instinctively chose for such poems a simple repeating metrical pattern, such as that which pervades the whole of 'Heimweh', where the open fifths in the bass and the naïve murmuring of the stream in the piano accompaniment, even the very characteristic repeating sequences, add to the song's unsophisticated air, which is only disturbed by the chromatic harmonies that so clearly suggest the puzzled unhappiness of the wanderer from home. This folk-song simplicity of form is often used by Mörike as a mask to conceal more complex sentiments. 'In der Frühe', for instance, opens with the poet lying restless and awake after a night of anxious insomnia and only comforted by the bells that begin to sound as day breaks. Wolf's setting follows in every detail the simplicity of the poem's form and the gradual change from tension to relaxation. The morning bells, when they rise sequentially, prove to be identical with the piano-figure which has dominated the first half of the song—lying low and harmonized chromatically at first, then rising into uninhibited euphony.

At the opposite extreme to the *Des Knaben Wunderhorn* element in Mörike's poetry lies the impassioned quasi-philosophical lyricism of the love-poems and letters. At the age of twenty the poet underwent a traumatic experience in his passion for Maria Meyer—a beautiful and certainly hysterical adventuress who won publicity by staging fainting-fits in strategically chosen public places, allowing herself to be 'rescued' and then disappearing after breaking a number of hearts, including Mörike's. He used this episode when he came to write his one novel *Maler Nolten*, where Maria Meyer appears as 'Peregrina' and is made the occasion of five lyrics inserted in the text—five of the most intense, immediate, and passionately expressed poems in the German language. 'As regards things of the mind,' wrote one of his friends, 'Mörike is like a son of Goethe's by some wild, mysterious marriage'—and the 'Peregrina' poems are the best illustration of this. Wolf set two of them, using for the second the full-blown Wagnerian language of chromatic suspensions and dramatically planned sequences that characterized the stricken, agonised erotic atmosphere of *Tristan*.

In complete reaction to the violent sensuality and disorder that Maria Meyer brought into Mörike's life, Luise Rau (to whom he was engaged for four years) was, he said, 'like an airy embodiment of my most sacred thoughts'. The poems which she inspired are in fact interspersed with images and expressions of religious devotion. In 'An die Geliebte' the poet finds his 'boldest, his unique desire' fulfilled in a love which unites the

sacred with the profane and sets the very stars singing—a
sentiment which Wolf echoes by a return to the world of
Tannhäuser.

With Mörike's directly devotional poems Wolf was not very
happy in either his choice or his handling. Both 'Karwoche' and
'Seufzer' belong too patently to the world of *Parsifal* and suffer
from rhythmic stiffness and monotony; 'Wo find ich Trost'
verges on the rhetorical and 'Neue Liebe' issues in one of Wolf's
most banal cadences, while his setting of the exquisite 'Gebet'
has been rudely compared to a harmonium voluntary in a village
church. The *memento mori* contained in 'Denk es, o Seele' on
the other hand, plainly stirred the composer. This exquisite
poem occurs in Mörike's supremely successful *Novelle—Mozart
auf der Reise nach Prag*—where it is referred to as a 'Bohemian
folk-song'. It has a note of eeriness that Wolf has perfectly
caught in the hesitant rhythms and suddenly broken phrases of
the accompaniment. 'There is a fir tree growing in the forest, a
rose blooming in some garden; remember, my soul, they are
fated to take root on your grave. Two black horses feeding in
the meadow, cantering gaily home. They will go at a funeral
pace when they draw your coffin—perhaps even before they cast
their shoes I now see gleaming.'

It was of course the duty of nineteenth-century critics to
discover in Mörike, as in all poets whom they admired, a
Hellenic strain. But in fact there was very little classical about
Mörike except the purity and clarity of the language that he uses
to describe or suggest even the most transient and mercurially
shifting sensations or states of mind. The Beauty which is the
conventional third person in his trinity 'proceeded' from the
other two, Nature and Love; and he was neither intellectually
nor aesthetically drawn to that neo-classicism which remains an
undercurrent in European art, sometimes disppearing for a
generation or more only to emerge in a new form. There was a
touch of the Walter Pater or even Oscar Wilde aesthete in one of
his stories—*Die Hand der Jezerte*—but the nearest approach to
this in his poems is to be found in those inspired by an antique,
'Auf eine Lampe'—or pictures—'Auf ein altes Bild' or 'Schlafendes Jesuskind'. These begin with a kind of Parnassian objectiv-
ity, very soon shot through with subjective, personal feeling. 'Auf
eine Christblume' begins with the apostrophe of a flower—the

Christmas rose—and by miraculously light and natural transitions displays almost the whole gamut of Mörike's intimate
emotional world—his feeling for the uniqueness of the flower,
its wintry whiteness as a symbol of purity, its purple spotted
heart marked with the signs of Christ's Passion. This, as it were,
classical vision is suddenly and most characteristically interrupted
by Mörike's picture of the elf—an unredeemed and irredeemably
Germanic figure—attracted by the radiance of the flower that
shines even in the darkness. He stands gazing a moment and
then flits uneasily away. There is none of Wolf's Mörike settings
that follows more faithfully or with greater aptness every
slightest image and reference, every infinitesimal change of
mood and tempo in the poem. From the hymn-like opening the
song proceeds like a well-composed picture in which every detail
is organic and images are skilfully correlated. The deer cropping
the snowy turf by the churchyard wall, for instance, is given the
same basic motif as the elf, the one conjured up by gently
repeated even quavers and the other by delicate semiquaver
triplets.

If the affinity between Mörike and Wolf is at first sight difficult
to understand, the truth is that Mörike is a good deal less simple
than he at first appears in many familiar poems and Wolf is a
good deal simpler. It is certainly true that those of his Mörike
settings that won most admiration for their complexity—those,
that is, in which Wagnerian harmony and Wagnerian psychology, as it were, were most remarkable—are no longer among
those which we most admire. It is not 'Seufzer' or 'An den
Schlaf' that we remember but 'Der Gärtner', 'Fussreise', or the
sublime simplicity of 'Schlafendes Jesuskind.'

24

GIACOMO LEOPARDI[*]

A poet is his language and his style; and the world of ideas and emotions from which he writes is more important for the character that it imparts to his language than for its own sake. The trivial, 'social' element in Byron's character, for instance, is of no interest or value in itself; but it accounts for the intimate, conversational tone that makes even the weakest passages of his narrative poems as enjoyable as the stories of a witty and well-bred raconteur. After nearly a century and a half we see much more clearly in the best of the so-called 'romantic' poets of the 1820s and 1830s the solid residue of eighteenth-century reasonableness, that counterbalanced their inclination to emotional and linguistic self-indulgence—and we see it particularly in their language.

Giacomo Leopardi was unique among romantic poets in being a classical scholar in whose collected works translations of Moschus, Virgil, Homer, Hesiod, Simonides, and the *Batrachomyomachia* occupy a larger place than the handful of canti, canzoni, odes, and hymns to which he owes his reputation. These begin in 1818, when the poet was twenty, with the classical ode 'All' Italia', and end with 'La ginestra', written in the spring of 1836, a few months before his death. In the anonymous preface to the 1824 Bologna edition of his poems Leopardi summed up the philosophy which had already been largely forced upon him by circumstances and was never to be modified:

After the discovery of America the world has seemed to us smaller than it did before. Nature spoke to—that is to say, inspired—the Ancients without revealing herself; and the more discoveries we make about the physical world, the more our imaginations are impressed by the nothingness of the

[*] BBC 20 December 1964.

whole universe. Everything is vain in the world except pain, and even pain is better than boredom. Our lives deserve nothing but our contempt. The inevitability of an evil will console commonplace minds for that evil's existence, but not great minds. Everything in the universe is a mystery except our own unhappiness.

This is a strange creed for a young man of twenty-six, even in 1824 when the cult of poetic melancholy was well established. Most of the poets of the age knew, and many no doubt deliberately cultivated, these moods of blank despair. Six years earlier for instance, Shelley had written:

> I could lie down like a tired child,
> And weep away the life of care
> Which I have borne and yet must bear,
> Till death like sleep might steal on me,
> And I might feel in the warm air
> My cheek grow cold, and hear the sea
> Breathe o'er my dying brain its last monotony.

> (from 'Stanzas written in dejection, near Naples')

In Shelley's case this was no more than a mood, but with Leopardi it was from an early age a settled emotional state, prompting him to adopt a philosophy of despair such as Shelley could never have imagined. Even in Leopardi's short lifetime—he died at the age of thirty-nine—he made his name as the poet of an exalted but intellectually argued melancholy, and there are in his poems as few interruptions of this mood as there are hints of its being anything but completely genuine and unaffected.

If we search the poems themselves for clues to the exact nature and possible cause of this melancholy, we shall not have far to look. At a very early age Leopardi alternates between lamenting a youth already lost, and declaring that he never had a proper youth. He was twenty-two when he accused Nature of creating him for a life of misery and denying him all hope for the future—this at his window on a summer night:

> . . . io questo ciel, che sí benigno
> Appare in vista, a salutar m'affaccio,
> E l'antica natura onnipossente,
> Che mi fece all'affanno. A te la speme
> Nego, mi disse, anche la speme; e d'altro
> Non brillin gli occhi tuoi se non di pianto.

(. . . I have come abroad now to salute
This sky whose aspect seems to be so gentle,
And ancient Nature powerful over all,
Who has fashioned me for trouble. 'I deny
All hope to you,' she has said, 'yea, even hope;
Your eyes shall not be bright for any cause,
Except with weeping.')

('La sera del dì di festa')

What in fact was this 'youth' that Leopardi was already
lamenting? and what hope had been denied him? Much later, in
1829, in one of his finest poems 'Le ricordanze' ('Memories') he
tells how 'even in the first youthful tumult of blisses, agonies,
and desires' he used to call on death, sitting for hours by the
fountain in the garden at Recanati, fascinated by the thought of
ending his hopes and his griefs in its waters. When he
apostrophises youth he reveals that he identifies it with love—a
girl's smile:

O primo entrar di giovinezza, o giorni
Vezzosi, inenarrabili, allor quando
Al rapito mortal primieramente
Sorridon le donzelle.

(Our opening time of youth!—Oh indescribable
And lovely days!—When first on the rapt mortal
Young girls begin to smile. . . .)

In the same poem he refers to himself as an 'inexperienced
lover'; and in 'Ultimo canto di Saffo' ('Sappho's last song'),
written as early as 1824, he makes almost the only allusion in all
the poems to the true root of his troubles:

. . . Oh cure, oh speme
De' più verd' anni! alle sembianze il Padre,
Alle amene sembianze eterno regno
Diè nelle genti; e per virili imprese,
Per dotta lira o canto,
Virtù non luce in disadorno ammanto.

(Oh! fear, oh! hopes
Of our first youth! it is to looks,
To charming looks that the Father of all
Gives eternal dominion among men; and despite a man's courage
Despite learning and the gift of song,
No excellence shines in an unlovely covering.)

That was Leopardi's tragedy—'an unlovely covering'. 'His figure was middling in height, slight and bent, his complexion pale,' wrote his friend Ranieri. 'His head was large, with a broad square forehead, languid blue eyes, a sharp nose, and fine-drawn features . . . his voice very low and faint.' The description is indeed a friend's and Ranieri loyally omits to say that Leopardi was in fact hunchbacked—a disaster in Italy, where this deformity has always been considered ridiculous. Indeed Leopardi could hardly have been born in a more unfortunate country than Italy where, even today (as Iris Origo tells us in her book on the poet): 'a man's physique is more considered than in the countries of the north; not only beauty and vigour, but a certain *allure*, a certain self-confidence and ease, are essential in order to please.' Leopardi never pleased. He fell desperately, humiliatingly in love with a succession of pretty, commonplace women who were faintly flattered and amused by his passion and his poems, but never for a moment took him seriously as a man. Whether owing to the general lack of self-confidence engendered by this unhappy circumstance or—as seems more probable—owing to some specific early experience, Leopardi tells us (in the autobiographical poem 'Consalvo') that the very idea of the physical consummation of love was always connected in his mind with an overwhelming fear:

> Sempre in quell'alma
> Era del gran desio stato più forte
> Un sovrano timor.

> (Always within that soul
> Stronger than his strong passion's self had reigned
> A sovreign fear.)[1]

It was no doubt because love, in the fullest sense, was unattainable that Leopardi came to identify it with all that is desirable in human existence—love whom he calls, as Dante had done, 'mighty master' ('prepotente signore'), though he never invoked Love as he does that other mighty force whom he regarded as the only alternative object of worship—Death.

Death he invokes again and again in language of almost

---

[1] This and the next translation, and those on pp. 288 and 290–1, are taken from G. L. Bickersteth, *The Poems of Leopardi* (Cambridge, 1923).

mystical reverence, but never in more passionate and eloquent words than in the poem entitled 'Amore e morte'. Love and death seemed to Leopardi twin divinities who between them command the only good things that man can hope for. When Leopardi wrote this poem he still had five more years to live; but he had already ruined by overwork and neglect the little health that his frail constitution had ever enjoyed. He turns to Death as to a power whom he has honoured and invoked for many years, one that owes him some return for his devotion in the face of vulgar hostility. Let Death, then, keep him waiting no longer:

> E tu, cui già dal cominciar degli anni
> Sempre onorata invoco,
> Bella Morte, pietosa
> Tu sola al mondo dei terreni affanni,
> Se celebrata mai
> Fosti da me, s'al tuo divino stato
> L'onte del volgo ingrato
> Ricompensar tentai,
> Non tardar più, t'inchina
> A disusati preghi,
> Chiudi alla luce omai
> Questi occhi tristi, o dell' età reina.

> (And thou, whom from the beginning of my years
> I have honoured and implored,
> O lovely Death, the one,
> The only comforter of human tears,
> If e'er I worshipped thee,
> If any insults at thy godhead hurled
> By the ungrateful world
> Were recompensed by me,
> Delay no longer, this
> Rare-prayed petition hear,
> Grant that these sad eyes be
> Now closed to the light, Queen of the centuries.)

The poem ends with a sudden burst of defiance, as unexpected as it is illogical—a refusal 'to bless the hand that scourges me and is dyed with my innocent blood', and an expression of aristocratic contempt for those who console themselves with such lies as are used to comfort children.

This familiarity with death, like much of his psychological

make-up, Leopardi doubtless learned or inherited from his mother. The portrait that he was to draw of her is a terrible one, for it is that of the first woman to refuse him love. We know that the Contessa Adelaide Leopardi was a masterful woman, married to an ineffectual husband who would have been financially ruined if she had not taken over the management of his affairs; and that she was intensely, obsessively religious, with a grim Jansenist piety that no Calvinist matriarch could have outdone. Several of her children died early and, according to her son, this was a matter of deep satisfaction to her:

When she saw the death of one of her infants approaching she experienced a deep happiness, which she attempted to conceal only from those who were likely to blame her; and the day of her child's death, if it came, was for her a happy and pleasant one, nor could she understand how her husband could be so foolish as to lament it. She considered beauty a true misfortune and, seeing her children ugly or deformed, gave thanks to God. In no way did she attempt to help them to hide their deficiencies . . . similarly she never neglected an opportunity of pointing out to them their faults. In these and other matters the shortcomings of her children were a comfort to her, and to talk of what she had heard said against them was always her deliberate choice.

Whether this is a true picture matters less to us than the fact that it is the impression that his mother left on the poet—nor is it difficult to imagine how such a woman might have set about instilling in a small impressionable boy that 'overwhelming fear' that was associated in Leopardi's mind with sexual pleasure.

Nevertheless I do not think there can be any doubt that it was from his mother that Leopardi, like many great men, inherited his temperament as well as his gifts. In fact the atheistic philosophy, which he horrified her by professing, in many ways resembled her own distorted religion—and in nothing more closely than in this obsession with death and in the insane pride of belonging to a spiritual élite which recognized the vanity of all earthly things.

Leopardi bitterly resented the suggestion that his pessimism was in any way to be accounted for by his poor health or unhappy circumstances. In fact the vigour with which he rejected this suggestion, when it was made in an article published in the German paper *Hesperus*, is itself suspicious and would suggest that the writer had indeed touched a tender spot,

far too near the truth for the poet's comfort. His reply, though
dignified, is not convincing:

Whatever my misfortunes may be . . . I have had enough courage not to
attempt to diminish their burden, either by frivolous hopes of a so-called
future and unknown happiness or by cowardly resignation. Before I die, I
must protest against this feeble and commonplace invention, and ask my
readers to attempt to refute my facts and my arguments rather than blame
my infirmities.

Until almost the end of his life Leopardi seems to have been
convinced that resignation was, as he says here, cowardly—
whereas of course it is the only attitude logically possible to a
nihilistic pessimist. Defiance of any kind was, after all, no more
than undignified shadow-boxing, as he came to see. No doubt,
though, the Contessa Adelaide had preached 'resignation in
affliction' and this was enough to make the very idea repellent to
her son. It is only in his last poem, 'La ginestra', that he came to
see the futility of protest against what seemed to him the
senseless unhappiness of human existence, the bitter contrast
between what life seems to promise in youth and what it
actually has to offer. He apostrophises the broom-plant growing
on the slopes of Vesuvius and doomed to extinction—yet not
resisting its doom, neither begging for mercy nor foolishly
challenging fate:

> Ma più saggia, ma tanto
> Meno inferma dell'uom, quanto le frali
> Tue stirpi non credesti
> O dal fato o da te fatte immortali.

> (But wiser still, and less
> Infirm in this than man, you do not think
> Your feeble stock immortal,
> Made so by destiny or by yourself.)

Much of Leopardi's worst misery consisted of what he
thought was a stoical, as opposed to Christian resignation, but
was in truth apathy—that *noia* or ennui that was made
fashionable by Byron and, when it was genuine, was almost
certainly due to directly physical causes. In a letter written in his
nineteenth year to his friend Giordani, Leopardi gives a vivid
description of this psychosomatic depression:

A few evenings ago, before going to bed, I opened my window and saw a clear sky and moonlight. With a mild breeze blowing and dogs barking in the distance, some of the old images awoke again in me and I felt my heart come to life again—so that I began to cry out like a madman, imploring some mercy from nature, whose voice I had heard after so long a silence. And at that moment, remembering my past condition (of apathy, that is to say), to which I was certain of returning immediately afterwards, I was frozen with horror—unable to understand how life can be borne without illusions or vivid affections, without imagination or enthusiasm.

Delicate, undernourished, overworked, psychologically un-stable, and so abysmally poor that he could only occasionally afford to leave the provincial palazzo where he lived with a father he could only despise and a mother whom he feared and hated, Leopardi might well have committed suicide. What in fact held him back? Pride, no doubt, as well as an instinct which had not been undermined by any examples that he would deign to imitate; and still, at this period, the desire for literary fame, though this was to prove the bitterest of all illusions. The most terrible of his poems, however, dates from much later, from the first weeks in which he finally realized that he would never know a happy love. 'A se stesso' ('To Himself') was written in 1833, the year in which his love for Fanny Targioni had reduced him to accept being tolerated by her for the sake of his handsome friend Ranieri, whom she did find interesting.

> Or poserai per sempre,
> Stanco mio cor. Perì l'inganno estremo,
> Ch'eterno mi credei. Perì. Ben sento,
> In noi di cari inganni
> Non che la speme, il desiderio è spento.
> Posa per sempre. Assai
> Palpitasti. Non val cosa nessuna
> I moti tuoi, nè di sospiri è degna
> La terra. Amaro e noia
> La vita, altro mai nulla; e fango è il mondo.
> T'acqueta omai. Dispera
> L'ultima volta. Al gener nostro il fato
> Non donò che il morire. Omai disprezza
> Te, la natura, il brutto
> Poter che, ascoso, a comun danno impera,
> E l'infinita vanità del tutto.

Iris Origo renders this faithfully:

> Now you shall rest for ever,
> weary my heart. Dead is the last illusion,
> that once I thought eternal. Dead. I know,
> for all the loved illusion of past days,
> not only hope, desire itself has fled.
> Rest now, yes, rest, for ever.
> Tremble no more, for nothing now is worth
> your restlessness and nothing worth your sighing.
> Life is bitterness and tedium—that and no more,
> the world a clod. Rest then. End your despair.
> One boon alone by fate
> to man was given—death. Now learn to scorn
> yourself, scorn Nature,
> scorn the brute power that, hidden, rules all for ill
> and scorn the infinite vanity of the universe.

This almost physical sense of misery, though nowhere else so strong, pervades practically everything that Leopardi wrote, except the prose works, where it is disguised by an oddly ponderous sarcasm. Yet it could never kill, and often serves only to intensify, his greatest poetic gifts—the purely musical effect of his language, which shines all the more brightly by contrast with the *idées noires* that it is used to express, and his feeling for natural beauty. Nature, as we have seen, appeared to Leopardi as a cruel mother—perhaps the Contessa Adelaide had made him think instinctively of all mothers as cruel. 'O courteous nature', he writes sarcastically in 'La quiete dopo la tempesta':

> O natura cortese,
> Son questi i doni tuoi,
> Questi i diletti sono
> Che tu porgi ai mortali. Uscir di pena
> È diletto fra noi.
> Pene tu spargi a larga mano; il duolo
> Spontaneo sorge . . .
>
> (O courteous Nature, these
> Thy gifts are, and 'tis thus
> That thou dispensest joy
> To mortal beings. Mere relief from pain
> Is counted joy by us.
> Pains dost thou strew with lavish hand; unbidden
> Sorrow arises . . .)

Despite this, the solitary gleams of happiness in Leopardi's poems are all connected with his spontaneous delight in natural beauty. The winter of 1827–8, which he spent at Pisa, was probably the most hopeful period of his life; and 'Il risorgimento', written the following spring, describes the struggle in him between the intellectual conviction of being destined to un-happiness and his physical and emotional delight in the reawakening of nature—the fountain speaks to his heart, he says, the sea holds converse with him:

> Se al ciel, s'ai verdi margini
> Ovunque il guardo mira,
> Tutto un dolor mi spira,
> Tutto un piacer mi dà.
>
> Meco ritorna a vivere
> La piaggia, il bosco, il monte;
> Parla al mio core il fonte,
> Meco favella il mar.
>
> (At heaven, at verdant river-brinks,
> Where'er I gaze in wonder,
> My heart with joy grows fonder,
> My heart with pain doth thrill.
>
> Once more I live in sympathy
> With fields and woods and mountains;
> The ocean waves, the fountains
> Once more are friends with me.)

Almost all Leopardi's greatest poems have a well defined natural setting. Many of the very finest are night-pieces—'Ultimo canto di Saffo', 'La sera del dì di festa', 'Alla luna', 'Le ricordanze', 'Canto notturno', 'Il tramonto della luna'. The fireflies hovering over the summer hedges, the cypresses stirred by a warm breeze; a still, cloudless night with the moon hanging huge over the olive groves and bathing every detail of the distant mountains in its placid light; the church clock striking the hours; a girl singing at her work in the lighted room of a lonely house—these passages once known recur again and again to the mind during an Italian summer. Or the superb description of the night sky over the Bay of Naples, with the glittering stars reflected in a sea as pure and untroubled as the sky itself:

In purissimo azzurro
Veggo dall'alto fiammeggiar le stelle,
Cui di lontan fa specchio
Il mare, e tutto di scintille in giro
Per lo voto seren brillare il mondo.

(Watch in pure azure skies
The constellations star by star emerging,
To which yon ocean lies
A distant mirror, till in calm profound
The world with sparks is glittering all around.)

('La ginestra')

The tense, distanced melancholy of these night-pieces and the
simplicity and purity of their language have often been
compared to Chopin's Nocturnes, though their beauty is more
severe and less elegant. Leopardi was too much of a classicist at
heart and too completely a Latin in temperament to fill his
landscapes with the closely observed physical detail of the Dutch
painters or the English poets. But there is one exception, in 'La
vita solitaria', and it is significantly not a night-scene but a
description of the trance-like stillness of noon in high summer
above a reedy lake—the sun reflected in the motionless surface
of the water, not a leaf stirring, no sound of a bird's flight, an
insect's hum, a cicada's chatter. The poet feels held like a fly in
this amber atmosphere, forgetting and forgotten by all, as
though tasting the dissolution of death in anticipation—and in
fact the passage comes between one of Leopardi's very rare
references to suicide and an agonised outburst in which he
complains that Love has been betrayed:

Talor m'assido in soliataria parte,
Sovra un rialto, al margine d'un lago
Di taciturne piante incoronato.
Ivi, quando il meriggio in ciel si volve,
La sua tranquilla imago il Sol dipinge,
Ed erba o foglia non si crolla al vento,
E non onda incresparsi, e non cicala
Strider, nè batter penna augello in ramo,
Nè farfalla ronzar, nè voce o moto
Da presso nè da lunge odi nè vedi.
Tien quelle rive altissima quiete;
Ond' io quasi me stesso e il mondo obblio

Sedendo immoto; e già mi par che sciolte
Giaccian le membra mie, nè spirto o senso
Più le commova, e lor quiete antica
Co' silenzi del loco si confonda.

(At times I seat me in a lonely spot,
Upon a gentle knoll, beside a lake
Ringed with a silent coronal of trees.
There, in the full noon of a summer's day,
The Sun his tranquil image loves to paint,
Nor grass, nor leaf stirs in the windless air,
No ripple of water, no cicada's shrill
Chirping, no flutter of wings upon the bough,
No buzz of insect, voice or movement none,
Far off or near, can ear or eye perceive.
Those shores a deep, unbroken stillness holds;
Whence, sitting motionless, I half forget
Myself, forget the world, my limbs appear
Already loosened, neither soul nor sense
Informs them more, and their age-old repose
Is mingled with the silences around.)

Recent references to Leopardi as a 'neurotic' poet seem singularly inept in view of this almost Virgilian sobriety of language and his unfailing ability to temper sensibility with sense. Of course by any standards of clinical normality at least eighty per cent of all poets must be classified as 'neurotic' but is it not, after all, only the severely 'disturbed' oyster that produces the pearl? The real test of any poet's fundamental sanity lies in the effect of his poems on the reader; and the strength and visionary beauty of Leopardi's language far outweigh the apparent darkness and restriction of his emotional world. There is, too, something enormously invigorating, exhilarating even, in the spectacle of a terribly handicapped man facing, and triumphing by sheer creative power over the very worst that even his imagination can represent as his fate—and this is precisely what Leopardi does in his greatest poems.

PART VII   RELIGIOUS THINKERS

25

GEORG CHRISTOPH LICHTENBERG[*]

A rod with a positive electric charge produces ray-like figures in iron-filings, while negatively charged it produces rounded figures—this simple discovery was made by an eighteenth-century amateur German physicist, Georg Chrisoph Lichtenberg, whose name has thus been modestly perpetuated in textbooks of physics. But Lichtenberg's interests were much wider than science or literature. Like the great French aphorists he was before all else a *Menschenkenner*—a student of human nature who sensibly turned his gaze inwards, on his own psychological organism, before generalizing from the observation of his neighbours, and planned an autobiography whose frankness should make Rousseau's seem pale. Instead of the autobiography he left a heap of 'scribble-books' which bear eloquent witness to what he described as his 'incredible pleasure in self-observation' and provide something like a portrait of the man himself. These 'Aphorisms' have never made such a reputation as those of the French moralists (perhaps owing to the fact that they are in German), but they have delighted such difficult connoisseurs as Schopenhauer and even Nietzsche, who numbered them among the five German books that he could always read again.

Lichtenberg was the eighteenth child of a country parson near Darmstadt, and a hunchback; but there is little trace in his writings or his life of the resentment and distortion that we—with our enlightened psychological views—should expect in a victim of poverty and physical deformity. Instead, the aphorisms show a quite unusually balanced personality, well developed sensually and emotionally as well as intellectually, a man of the eighteenth-century enlightenment who nevertheless recognized the limitations of human reason and insisted a good century before Freud that dreams provide an essential key to the

[*] BBC 16 August 1964.

understanding of human character—'if people told their dreams honestly, these would be a better guide to their character than are their faces.' Nor can I think of any other writer who ranges so freely and naturally from frank and witty discussion of sensual pleasure to a reasoned defence of Jacob Boehme's mysticism, displaying over the whole journey a fantasy and a freedom from dogmatism of any kind that are both rare in the eighteenth century. He interrupts his aspiration for a Christianity 'freed of clerical nonsense' with the reflection that 'what is to guide men must be indeed true, but it must also be universally intelligible—even if imparted in pictures which humanity explains differently to itself at different levels.' 'Read this man's writings,' he said of Boehme, 'and deny their inner meaning if you can. We laugh at Boehme, as though the supernatural that he wanted to express could sound natural. Would an inhabitant of Mercury recounting in German the observations of other senses than ours sound more rational?'

Lichtenberg went as a student to Göttingen, which he made his home for the rest of his life, living in the house of Johann Christian Dieterich, founder of the Almanach de Gotha, and spending the last twenty years of his life—from 1778 to 1798— as editor of the *Göttingen Taschenkalender*. But beneath the witty and polite man of the world, honoured by foreign academies and sought out by visiting notabilities, lay the private man of feeling who enjoyed contemplating death and suffered from depression. He described himself as 'a pathological egoist' and at the age of thirty-five entered on a passionate relationship with a flower-girl of thirteen with whom he spent four years of 'labyrinthine oblivion' and half-concealment until her death came as the emotional turning-point in his life—after which he married happily and bred a family of nine children before dying in 1799, at the age of fifty-seven.

In Lichtenberg the characteristic scepticism of the eighteenth-century intellectual is certainly fundamental, yet it is perpetually corrected, not only by the man's sheer warmth and generosity of feeling but by the radical nature of his intelligence. Although we find him naïvely tracing every evil in the world to the 'often unconscious respect for old customs and old religion', he knew very well that 'disbelief in one thing generally means blind belief in another', and he was no naïve rationalist. 'Reasons,' he observed, 'are for the most part an attempt to give an appearance of justification and rationality to what we in any case intend to do.' Belief in God, however difficult to formulate intellectually, is for him 'an instinct—one that can be modified or even suppressed but is normally there and indispensable to a man's proper inner formation.' The study of Nature and Philosophy appeared to him to destroy the belief not in God, but in the 'succouring' God of our childhood. God appears rather to him as a Being 'whose ways and thoughts are not ours'. Was he knowingly quoting Isaiah? If so, it is strange that he concludes that such a God—whom on another occasion he can even define as 'the personification of incomprehensibility'—is of no assistance to helpless man. The truth is that Lichtenberg returned again and again, with refreshing honesty, to the fundamental questions of religion, approaching them from different angles and finding a variety of answers but never resolving finally (as only a superficial thinker can claim to do) the opposition between the mechanical conclusions of pure ratiocination and 'arguments' prompted by the whole personality. *Erst müssen wir glauben, dann glauben wir*, as he put it—in the first place we *have* to believe, and then comes belief.

Nature was for him, as for all his contemporaries, the great teacher. Men are her pupils, but unfortunately spend less of their time listening to her than to the chatter of their fellow-pupils, whose notes they crib without really understanding them. He faced the human constitution with admirable good sense and humour. 'The body is built on three floors—if only the tenants of the top and the ground floors could get on better!' Denial of man's instinctive proclivities is unrealistic and educationalists should use the illogical rather than deny it. 'If I say, "wash your teeth every morning", I am less likely to be obeyed than if I say, "take your two middle fingers and make the sign of the cross"—well, we should take this too into account. Heaven preserve us from turning men, who should have all Nature as their masters, into mere wax-lumps bearing the likeness of some high-minded professor.'

Mere accomplishment, intellectual or otherwise, could not greatly impress a man of learning who observed with distaste of someone that 'the man is so intelligent that he is good for nothing'; nor could it fox the gifted linguist who hit upon the comforting truth that 'to speak a foreign language perfectly and with the correct accent one must not only have a memory and an ear, but also be to some extent a little fop'. Some of his educational principles were far ahead of his time and have not really been accepted yet. Dogs, he thought, should be brought up with children rather than with intelligent adults, just as children should be kept with people only a little more advanced intellectually than themselves—an unexpected bouquet for the old nursery upbringing.

He was impatient of an educational system that attached more importance to the study of Roman history, and even mythology, than to science, and left men in almost complete ignorance of their own bodies. 'Get to know your own body and what you can of your soul,' he advises. 'Sharpen your intelligence by practical geometry (by which he meant surveying). But'—with his eye on the merely learned—'avoid an encyclo-paedic knowledge of worms—something that is useless if incomplete and all-consuming if complete. I grant that God is as eternal in the insect as in the sun. But He is also immeasurable in the sands of the seashore, which not even a Linnaeus has attempted to catalogue.' During a period of depression towards the end of his life Lichtenberg was tempted to despair of all human knowledge except mathematics. The sense of being unable to penetrate nature's secrets almost led him to believe, he tells us, that the sea-shells found in mountain regions really originated there and provided no firm evidence of geological upheavals. But a few days before his death, discussing the nature of mathematical genius with one of his sons, he had regained his confidence. Irrefutable intelligibility, he said, such as is to be found in mathematical propositions, is the equivalent in the spiritual sphere of digestibility in the physical. Men might live for ever if they could support existence on such a bland diet. What a contrast with the human reality, as Lichtenberg pictures it in an almost Shakespearean tirade: 'Man says—"I am made in God's image"—and drinks the urine of the immortal Lama . . . kills himself, declares himself divine, castrates himself, burns or whores himself to death, takes vows of chastity and burns Troy for a trollop . . . eats his fellow men and his own excrement.'

We repeatedly find him breaking the neat and arid philo-sophical moulds of his day in such flights of imagination. 'What

I dislike about our definitions of genius,' he said, 'is that they contain nothing about the Last Trump, nothing about "re-sounding through eternity" and "the footsteps of the Almighty".' Or he will throw a surprisingly long glance into the future—'if the sexes were not distinguished by their dress but had to be guessed, we should have a whole new erotic world, something worth making the subject of a novel by a writer who combined wisdom and knowledge of the world.'

England played an important part in Lichtenberg's intellectual and emotional life, as it did in the imagination of so many of his contemporaries—Voltaire, Montesquieu, Goethe, Pombal. He learned his English, which David Garrick himself was later to praise as unique in his experience of foreigners, among the young Englishmen who were sent to Göttingen to learn German; and in 1774 he paid a visit of eighteen months to London, where he boasted of having lived the life both of a workman and of a lord—he was in fact for some time a guest of Lord Boston's. George III had given him astronomical work to do in Hanover and visited him often at Kew, where Lichtenberg worked at the Observatory. He interested himself in the English theatre of the day and held strong views on the comparative worth of Sterne, whom he enjoyed but held rather small, and Fielding, for whom he had great admiration. His English notebooks contain an account of one of those unforgettable and inexplicable emotional experiences that are familiar to very many men of feeling and seem always to take place in a foreign country. On 15 April 1775, 'being the day before Easter Sunday', Lichtenberg went for a walk in Hyde Park at six-forty-five in the evening—the day and hour are noted as exactly as in Pascal's great document. The solemnity of the day and the sight of the Paschal full moon over Westminster Abbey filled him with a heightened emotional awareness as he wandered down Piccadilly and the Haymarket to Whitehall. Here he fell in with an organ-grinder, whom he followed as far as the Banqueting House, where the man suddenly began to play Lichtenberg's favourite chorale, 'In allen meinen Taten'; and the combination of the familiar tune and all its childhood association, with the thought of Charles I walking to his death from the window above him, stirred in Lichtenberg one of those complex emotions compounded of nostalgia and a sense of euphoria, in which his pains and worries suddenly became sovereignly unimportant. 'I could not help singing quietly to myself the words, "Hast Du es dann beschlossen, so will ich unverdrossen an mein Verhängnis gehn." ' ('If Thou hast so decreed, I will not wait nor plead, but go to meet my end.') Lichtenberg knew himself well when he wrote, 'When my spirit takes wings, my body falls on its knees.'

In fact the whole gamut of human feeling was familiar to this little, bright-eyed hunchback, whose body demanded pleasure as imperiously as his mind demanded activity. He was aware, sometimes painfully, of being precariously suspended between intellectual or spiritual contemplation and uninhibited sensuality. 'My body and I have never been so separate as we are at present,' he confides on one occasion to his notebooks. 'Sometimes we don't recognize each other at all, and then we collide so sharply that we don't know where we are.' Lichtenberg was a perfect example of Nietzsche's observation that 'the nature and degree of a man's sexuality reach to the furthest recesses of his personality'. His wit and his linguistic virtuosity— both closely connected with his sexual vitality—give many of his aphorisms a harmless, even frivolous look that did not deceive Goethe. 'Wherever Lichtenberg makes a joke,' he said, 'you will find some problem hidden.' This is particularly true of his apparently casual comments on human nature. 'Nothing reveals a man's character more clearly than a joke that he takes badly,' is one of many Freudian pre-echoes. 'If only I could make up my mind to be well', is another; and there is food for a whole reflective essay in the deceptively simple observation that 'a shameless man can look modest if he wants to, but a modest man cannot look shameless'.

As soon as Lichtenberg comes to politics or sociology he is far too intelligent to fall into the nets of any school or party. He had of course the sympathies and the special insights shared by all intelligent men of his day. His contempt of royal persons as such was founded not on prejudice, but on the evidence of his own eyes. His list of the misdemeanours or psychological peculiarities of his royal contemporaries makes very funny reading. It begins, 'the King of France makes pies and corrupts decent girls . . . the last King of Poland used to shoot at his court-fool's backside with a blowpipe . . .' and ends, 'the Duke of Württemberg is a lunatic, the King of England makes whores of Englishwomen, and the Prince of Weilburg bathes publicly in the River Lahn.' As a rider to this he tells the cautionary tale of the wig-maker who delivered himself in a Landau taproom of the pious wish that the 'world would wake up and send three million to the gallows, so that perhaps fifty to eighty million might be made happy'. He was arrested, observes Lichtenberg coolly, as mad ('quite rightly'), and before reaching prison he was beaten to death by a sergeant, who was himself beheaded.

Lichtenberg was guarded in his admiration of the French Revolution, being sadly aware that 'the highest degree of political liberty is next door to despotism', and that 'the equality we demand is in fact the most tolerable degree of inequality'. He saw clearly that the excesses of the French Revolution inevitably caused 'every intelligent demand justified by human and divine justice to be regarded as a germ of revolution'. He was looking a long way into the society of the future when he asked whether, 'when we break a murderer on the wheel, are we not making the same mistake as the child who hits the chair on which he has knocked himself?'

The aphorisms introduce us to something more than a new mind and a fresh kind of sensibility and humour (who would guess that any man of the eighteenth century could say that a donkey 'looked to him like a horse translated into Dutch'?); but in them we also meet a new friend—which is hardly the way one would describe La Rochefoucauld and the French aphorists. What Lichtenberg himself felt about the Psalms and their authors, a modern reader can feel about Lichtenberg himself:

I read the Psalms of David with great pleasure, because they show me a great man often feeling exactly as I do myself. When he gives thanks for delivery from his afflictions, I feel 'one day the time will come when you too will be able to give thanks for such a delivery'. It is a real consolation to see that a great man in a higher position than one's own was no happier or contented than one is oneself, and that after thousands of years he is still remembered as a source of consolation.

26

FRANÇOIS DE SALIGNAC
DE LA MOTHE FÉNELON*

Our grandparents thought of Fénelon as a man of goodwill unjustly persecuted by an overbearing rival—the dove of Cambrai, as it were, in the claws of the eagle of Meaux, a character all sweetness and light, a precursor of the tolerant and enlightened eighteenth century silenced by the bigotry and absolutism of the Grand Siècle. He owed this reputation partly to the fact that he came into conflict with Papal authority over his championship of Mme Guyon's mystical doctrine; and partly to the tone of his *Télémaque*, the only one of his writings to be well known in this country and a book in which a horror of despotism and of the spirit of ownership does indeed look forward at least a hundred years.

During the last forty years, however, Fénelon's reputation, like that of many other historical characters, has entirely changed. The gentle dove has become a serpent; the apostle of tolerance has been unmasked as permitting, if not instigating, the forcible 'conversion' of his Protestant fellow-countrymen; and his momentary unorthodoxy has come to seem a very small thing beside the complete, unquestioning submission which was his reply to Papal censure. The charges that Kingsley made against Newman have been revived against Fénelon, and he has been depicted as tortuous in intrigue, a master of equivocation in controversy, and something less than manly in temperament—almost the sinister priest-figure of popular imagination.

None of these ideas of Fénelon is in fact without some foundation, for he was a complex and in some ways contradictory character—an Ultramontanist and a mystic, an intellectual in love with spiritual simplicity, and a courtier who

* BBC 22 September 1964.

interested himself passionately in the organization and well-being of his rural diocese.

The key to the man is to be found, I believe, in his informal letters of spiritual advice rather than in his formal works. In the letters he sometimes openly, more often implicitly, reveals his true character and preoccupations. These are very far from consisting entirely, or even chiefly, in esoteric discussions of the mystical doctrine of 'Pure Love' with which his name became linked after the *cause célèbre* with Bossuet. But it is nevertheless impossible to ignore this whole central incident in Fénelon's life, if only because his encounter with Mme Guyon and her teaching formed, as far as we can see, the starting-point of his spiritual development; and the suffering and humiliation involved in his public championing of her cause against Bossuet gave Fénelon's mature character and teaching its strength, balance, and simplicity.

When he met Mme Guyon, in 1688, he was already thirty-seven, a well-connected and successful young prelate, who was nevertheless painfully aware of being the victim, not of any obvious vice but of that thoroughly respectable *amour propre*, or self-love, that is the obstacle to all true spirituality. His reputation in the world, his friendships, his quiet pleasures, and modest comforts formed his horizon. 'The whole root of our troubles,' he was to write later, 'is that we love ourselves to the point of idolatry, and everything that we love outside ourselves, we love only on our own account.' Mmme Guyon's doctrine of pure, disinterested love came to him not so much as a revelation—the 'amalgamation of two sublimes', as Saint-Simon caustically described it—but as a badly needed medicine.

The essence of Christianity, she taught (and she was merely rediscovering an age-old doctrine that had been temporarily obscured), is the soul's reception of God's gift of love—the ability, that is to say, to love not selfishly, but as He loves. This demands an act of disappropriation, a disowning of the self, the abolition of the active egocentric will and the substitution of the passive, perfectly detached will centred on God. (How far this passivity can or should extend, and whether it can become a habitual state rather than a series of repeated acts, were the chief points at issue in the dispute between Fénelon and Bossuet, but they do not concern us here.) The proof that, whatever Fénelon may have been supposed to say in his formal theological writings, he did not in fact believe or counsel anything other than a newly 'interiorized' Christian spirituality, is to be found in his letters. Anything less like the exaggerations attributed to Mme Guyon it would be hard to imagine. She had been a necessary stage on Fénelon's way, accepted rather than welcomed; he seems indeed to have been embarrassed, even repelled, by her gushing manner, her private jokes and pet names, which automatically gave an air of coquetry to a relationship which he envisaged more as that between teacher and pupil. Mme Guyon's sole importance for him lay in the fact that she had experienced what was to him only a theoretical relationship with God as Love, and could to some extent communicate that experience. In his subsequent personal defence of her under official attack there was perhaps as much chivalry as personal feeling.

At the heart of Fénelon's spirituality lie four qualities—simplicity, intimacy, gaiety, and peace (*une paix sèche*, a dry, unemotional peace, he called it). To realize how unconventional, almost revolutionary this composite ideal could appear at the end of the seventeenth century in France, we have only to glance at the manuals of piety, the sermons, and the theological treatises of the day, with their formal and lengthy devotions, well-argued points, and courtly baroque style. If Fénelon's aim is that truly supernatural *désoccupation de soi*—the gradual recentring of the whole, naturally egocentric, nature upon God—the means he uses to promote this are sublimely simple and commonsensical. 'Do not fight or deny your temperament,' he says, 'but accept it and learn to use it.' For 'it will never be your temperament that God will hold against you, since you did not choose it and are not in a position to alter it. If you bear it as a mortification, it will even contribute to your sanctification.'

This was revolutionary advice, quite contrary to much of the moral theology of the day, which too often held up stereotyped models of perfection to be copied and did not take into consideration the infinite diversity of human nature, the impossibility of forcing all characters into the same mould.

Fénelon valued the joyful acceptance of the incidental 'mortifications' or small contrarieties of everyday life far more highly than any formal, deliberately chosen mortifications,

which he mistrusted as too often fostering that self-complacency
that was always to him the supreme enemy:

Cultivate before all else a spirit of peace. Be free, gay, simple, like a child—
but a bold child, who is afraid of nothing, says everything frankly, and
allows himself to be guided or carried—a child that knows nothing,
foresees and plans nothing in advance but enjoys a liberty and a boldness
unattainable to grown-up people.

Here he is anticipating St Thérèse of Lisieux and his ideal, like
hers, involves—for all its sweetness and light—austerities that
may well alarm the natural man quite as much as the
flagellations and so forth generally connected with the word
'mortification'. For Fénelon really believed in the necessity of
stripping away all the attachments by which the self clings to its
own flattering image:

I am only aware of my hair when it is torn from my head. God gradually
develops the foundation of our personality, which is at first unknown to
us; and we are astonished to discover, rooted in our very virtues, vices of
which we had always thought ourselves incapable. What we have to part
with are no doubt trifles, but trifles which form the very essence of our self-
esteem. On such occasions one would a hundred times rather fast on bread
and water all one's life, and practise the greatest austerities, than bear what
goes on within one.

The instance that he gives of what he means by such 'trifling'
mortifications of self-love are revealing, for they have a strongly
personal flavour:

We are obliged first to talk naïvely, then to remain silent; to be first
praised, then blamed, then ignored, then re-examined. We are asked first to
stand down, then to stand up—to allow ourselves to be condemned
without saying a word on our own behalf, and then on the next occasion to
defend ourselves. We have to face finding ourselves weak, anxious, worried
by trifles; showing pique like a child; shocking our friends by our lack of
feeling; becoming jealous and mistrustful for no reason. We have to learn
to speak patiently and frankly to those who do not inspire these feelings in
us and themselves resent such feelings; to appear unnatural and insincere,
and finally to find ourselves arid and low-spirited, with no taste for God,
our thoughts distracted and so far removed from any feeling of grace that
we are tempted to fall into despair.

There can, I think, be no doubt that we are reading here a
page of autobiography, a detailed factual account from Fénelon's

point of view of the agonies through which he was made to pass during the long-drawn-out, highly publicized dispute with Bossuet. This began in 1694 and only ended five years later with Fénelon's public discomfiture and withdrawal. We have seen what discomfiture meant subjectively to Fénelon himself. Historically the triumph of Bossuet's well-meaning but narrow-minded authoritarianism entailed nothing less than the eventual secularisation of humanism in France—the eighteenth-century gulf between the Church and the apostles of social, political, and educational progress. Fénelon could have bridged the gulf, which was still small in his day, and he had instead to watch the ruin of his hopes and the spectacle of his own humiliation.

But he was right in thinking that what had happened to him on a large scale—on a national stage and in the full light of publicity—happens to every human being on some scale or another. The thwarting or disappointment of our own ideas and ideals, which he likens to having the hairs plucked one by one from our heads, is a universal human experience from which we can all learn. And the only way to learn is by cultivating *souplesse*—malleability. God does not demand the sacrifice of the evil in our characters only; if we let Him, he makes it possible for us to sacrifice things in themselves harmless, even good, for the sake of the better and the best. This impulse to give without thought of reward is the lover's instinct, which Fénelon tried to instil and foster in all those who asked his advice.

Much of the spirituality of the day was formal, even actuarial in tone—calculating probabilities and drawing up, as it were, insurance policies on a minimum premium basis, approaching 'salvation' in the spirit of a business undertaking governed by the intricacies of commercial law. In protest against this spirit of enlightened self-interest, Fénelon demands in his correspondents the generosity of the lover. Yet there is nothing emotional in his 'pure love'—'a love which loves without feeling, just as pure faith believes without seeing', he calls it. It is centred in the will and brings with it no sensible consolations. Fervour is a quite irrelevant element in prayer—no more relevant than, say, the weather, which may be warm or cold, fine or stormy. In fact the genuineness of the soul's love of God is only really tested when his 'weather' is poor, in those times of 'dryness' which Fénelon himself knew so well and came to value as precisely the most valuable periods of testing and growth: 'The whole of religion consists simply of getting outside oneself and one's self-love and centring one's life on God. Prayer is never so pure as when one is tempted to believe that one is not praying at all.'

In case this may sound either unreal or impossibly heroic, Fénelon insists that the smallest and most humdrum events of everyday life in the world provide precisely the evidence of God's Will for each one of us, and the material He has designed for our sanctification. We are not to look elsewhere for occasions to practise virtues more to our own taste, but to learn to accept, as unquestioningly as small children, the rough and the smooth of everyday life. When things go badly and we suffer: '. . . it is enough to make ourselves small and unresisting. This is not courage, but something both less and more than courage—a weakness that recognises our own lack of strength and God's all-powerfullness.' Later in the same letter he writes with a penetration that certainly comes from personal experience:

Sometimes one suffers without knowing exactly whether it is real suffering or not. At other times the suffering is real enough, but one is aware of bearing it badly, and the consciousness of one's own impatience is like a second cross, heavier than the first. . . . Then one can be really happy, for a cross is no longer a cross when there is no longer a self to suffer its weight—a *me* claiming as mine all advantages and set-backs.

Fénelon numbered among his correspondents many courtiers and *grandes dames*, including Mme de Maintenon. He always insists they should carry out faithfully the duties of their positions, and warns them against making pious practices an excuse for neglecting these. Faults and imperfections of character worry him little compared with any hint of self-sufficiency, spiritual pride, or Pharisaism. Towards these he is merciless, and we find him writing to the Duchesse de Noailles:

I should be less worried to see you fault-finding, touchy, short-tempered, uncontrolled, and then thoroughly ashamed of yourself, than to see you perfectly regular in all your duties and in every way irreprehensible, but at the same time fastidious, haughty, cold, stiff, easily shocked, and with a good opinion of yourself.

He had no hesitation in rebuking Colbert, the Archbishop of Rouen, for his passion for building:

Building becomes a passion like gambling and a house can become like a mistress. Have you no more urgent use for your money? Remember, My Lord, that your revenues are the patrimony of the poor, and that these same poor are dying of hunger on all sides.

On at least one occasion Fénelon carried his criticisms very much higher than a local archbishop. An Appendix found in the library of Cardinal Fabroni at Pistoia contains his ideas on the reforms needed in the Papal Court, which he describes as 'motivated by vanity and political considerations, haughty in its dealings with the weak and almost shamefully accommodating in its dealings with the powerful'. All reforms, he insisted, must come from within and begin with a profoundly humble admission of guilt.

If 'joy' is the key-word of Pascal's spirituality, 'peace' is Fénelon's. The difficulties of his own temperament are always the determining factor in a writer's ideas; and Fénelon, as Père Varillon says, knew well that sense of despair that can attack 'a man of feeling who had little sensible experience of God and an intellectual who saw unusually clearly the physical elements in all intelligence'. Fénelon is one of the least carnal of French writers—Ernest Renan was to follow him in this—and the absence of physical 'sap' determined not only his style, which is airy, graceful, and insubstantial, but the character of his thought. Like Newman, Fénelon was not built for the robust spirituality which issues in a joy almost as physically infectious as laughter; and like Newman he has had his temperament unjustly held against him. In fact he speaks to perhaps an increasing number of modern readers by virtue of his acute intellectual and psychological awareness, his extraordinarily 'modern' simplicity and avoidance of pious jargon, and by that good-humoured, very French realism that is invaluable, though, by no means common, in spiritual writers.

27

FRIEDRICH VON HÜGEL*

Friedrich von Hügel was of 'mixed stock', as we call it; and it is worth while recalling something of these crossings, tensions, and border-line disputes in his actual physical origins, because they contributed to a character whose chief greatness lay precisely in an heroically achieved reconciliation of opposites and a resolute refusal to be contented with any one-sided view of life.

His father's family came from one of the great border-line districts of Europe—the Rhineland—and moved to another—Austria, in whose diplomatic service he ended his career. Friedrich von Hügel himself was born in Florence in 1852, and spent his childhood in Brussels, where his education was undertaken by a Protestant tutor under Catholic supervision. His mother was a Scot; and although he came to England as hardly more than a boy and married very young an English wife, he remained to the end of his life more naturally at ease with his mother's countrymen—especially the Scots philosophers and moralists—than with the empirical, lazy-minded, semi-Pelagian English, who, he complained, 'have not had the boy knocked out of them, the boy who dares not come out of his skin, the boy who is ashamed even to be accused of thinking'. 'It is, I suppose, English,' he added ruefully, 'but slipshod, go-as-you-please—a poor kind of Englishness.'

The extraordinarily wide and complete European background was matched by a corresponding breadth of intellectual interest: historical, philosophical, religious and even, in a rudimentary way, scientific. On the other hand, he was always handicapped by an extremely sensitive nervous system and a doubtless consequent inclination to go to extremes: he once described himself as having a 'very vehement, violent, over-impressionable

* BBC 7 December 1963.

nature'. He was deeply aware of this danger—'so glad you don't strain,' he wrote to his young wife in 1884, *'never* strain'—and twenty-five years later he observed to his niece Gwendolen Greene that 'the soul's health and happiness depend upon a maximum of zest and as little as possible of excitement . . . Zest is the pleasure that comes from thoughts, occupations, etc., that fit into our extant habits and interests, duties and joys that give us balance and centrality. Excitement is the pleasure which comes from breaking loose, from fragmentariness. Zest is natural warmth, excitement, fever heat . . . whose sirocco air dries up our spiritual sap.'

Although he was so alive to the danger of strain, the very vocabulary of von Hügel's letters is enough to show how he valued above all else commitment and the readiness for sacrifice. The neglect of his writings today is no doubt partly due to his handling of the English language, in which German syntax often combines with a determination to leave no shade of his meaning unexpressed, to create the most uncouth phrases and the most formidable sentences. Yet no one can dispute either the personal quality of his writing or the fact that it often furnishes a key to the character of his deepest aspirations. 'Bracing' and 'costing' (even the noun 'costingness') are among his favourite words, showing his impatience of all facility and superficiality; while his very personal use of the word 'utter', to express the limitlessness of God's attributes and what should be man's response to them, reveals what was the core of his existence.

Brought up in a conventional Catholic family, he experienced two major accessions or intensifications of his religious faith— what in fact an earlier generation would have called conver-sions—the one at eighteen, when an attack of typhoid left his hearing seriously impaired for life; and the other fourteen years later when, as a young married man in Paris, he came under the influence of the Abbé Huvelin. Although he never abandoned his general historical, philosophical, and literary interests, he had already in the 1880s begun to concentrate his attention on problems of biblical scholarship and it was to his slightly later interest in the mystical element in religion that we owe the greatest of his books. Mysticism he in fact regarded as 'the mature man's approach to religion and the foundation of true humanism'.

Now biblical scholarship and mysticism were dangerous interests for Catholics during the later years of the pontificate of Leo XIII and the whole pontificate of Pius X. They were dangerous because they were the chief fields to which the so-called 'Modernists' applied the theories or systems that ended by taking several of these men—among whom were many priests—outside not only the Church but any theistic faith whatever. Their wide influence and eventual defection caused a sudden and disastrous intensification of that 'siege' or 'ghetto' mentality that had afflicted Catholicism ever since the Counter-Reformation and issued in a number of sternly repressive measures.

Von Hügel numbered the chief Modernists, both French and Italian, among his closest friends. He frequently visited and corresponded with Loisy, Hébert, and Laberthonnière in France, Semeria and Fogazzaro in Italy; while the English Jesuit, Father Tyrrell, who was obliged to leave the Society and was suspended from his priestly functions, was probably the dearest and closest of all his friends. The 'terrible years' of the Modernist controversy—roughly between 1894 and 1914—formed in fact the central and critical period of von Hügel's life, during which he was incubating in his own person, by painful experience as well as deep thought, what were to be his mature attitudes to the perennial problems that concern Christians in a special context, but ultimately touch all thinking men and women.

How many of these problems in fact involve the reconciliation of two apparently contradictory forces or principles! Liberty and authority, intellect and intuition, human suffering and an all-powerful and loving God. And von Hügel's greatness as a religious thinker lay just precisely in this difficult or (as he himself would say) 'costing' reconciliation—in the refusal to allow one good to be sacrificed to another, in the insistence on not shirking the painful, long-term task of saving *both* goods for a larger and more 'central' synthesis.

The worst of the Modernist controversy, in which he had taken an active part and which incurred grave official displeasure and risked disciplinary measures being taken against him, was over in 1914. But by then Europe was locked in the First Great War and von Hügel had only ten more years to live. These ten years however were to prove incomparably the richest period of his life. He had always thought of himself as a 'thinker's

thinker', and his great studies of Mysticism and Eternal Life
gave him every right to do so. Now he found himself, at over
sixty, in ever-increasing demand as a lecturer—often to audiences
by no means learned—as an apologist, and most of all as a
spiritual adviser. He expressed his astonishment at this turn of
events in a characteristic phrase when he said that he found
himself 'awarded the first prize among cats, when I only
expected to do reasonably well among dogs'.

There is certainly great interest for the historian and bio-
grapher in his earlier correspondence with scholar-friends and
fellow fighters in the cause of liberty of research and expression
for Catholic thinkers; but it is in what may be called his
'spiritual letters' that we see the full stature of the man himself
and the extraordinary strength, originality, and 'central sanity'
(as he might well have called it himself) of his position.

It has been well said that Modernism played the same part in
von Hügel's development as Quietism in Fénelon's and Jansen-
ism in Pascal's. The stresses and heart-searchings caused by
these movements produced a kind of exterior discipline, intellectual and psychological, through which these profoundly interior
souls had to pass in order to achieve their full depth and
maturity. Von Hügel himself had an inkling of this, when he
compared the role of the natural sciences today—which he
whole-heartedly welcomed—to the 'humbling, purifying *factual*
element' which religion always needs, to prevent it from
becoming overweening in its demands and insisting on a quite
improper domination of other autonomous fields. Writing as
early as 1900 to Maud Petre, he praises a new book, by the
Protestant theologian Eucken, whom he deeply admired, and he
singles out:

his vivid consciousness of how the character and personality, the spiritual
substance of the soul, have to be won and conquered through constant
effort, renunciation, conversion, and purification; of how selfish and self-
centred, how animal and sensually sentimental is the natural man; and how
his childishness has ever to be turned to childlikeness, and even his
apparently good aspirations be thwarted and broken, so as to grow in
worth and range. All this would actually make him seek and postulate, in
such a moral training-school, just precisely the friction, the non-fit, the
otherness of Science and of religion—just exactly the scheme of things in
the midst of which we are. In the foreground ourselves, selfish, sensual

childish individuals, *mere units* but with the mysterious capacity (not more!) for constituting ourselves unselfish, spiritual, manly personalities, real unities and organisms. In the middle distance the phenomenal curtain and as it were, buffer-state, the resistant but spiritually not irresistible medium of the world of physical, mechanical, determinist, fact, law, and science; in the background, which is really the groundwork also of all, the noumenal reality, the world of spirits and of the absolute Spirit, of persons and the absolute Person, the world of morality, eternity and love.

Von Hügel believed passionately in the importance to any fully balanced religion of three elements—the intellectual, the mystical, and the institutional. 'Religious development,' he once wrote, 'means adding speculation to institutionalism, and mysticism to both.' And it was because Roman Catholicism had, he believed, preserved (however precariously and, at times lop-sidedly) all three elements that he remained, through all the gravest difficulties and doubts of the Modernist crisis, a loyal Catholic, extraordinarily simple in his practice however complex in theoretical allegiance. He set very little store by mere 'uncosting' cleverness but proclaimed instead 'the soul's humble, faithful, loving search for material, objective truth by an ever-growing purity of disposition and intention, and an ever-increasing attempt to become, and be, all it knows.'

In a letter to Algar Thorold, written in 1921, he contrasted two types of mind. The one, he says, sees truth as so many geometrical figures within which all is safe and correct, while outside lie danger and incorrectness. (These we may, I think, safely infer were the minds of the Vatican authorities and the nervous conservative Catholics in all countries, from whose incomprehension he suffered all his life.) The other type of mind, which is his own, sees truths as 'intense luminous centres' with a semi-illuminated outer margin, and then another and another, until all shades off into darkness. Such minds, he adds, are not in the least perturbed by having to stammer and to stumble. When they have moved out some distance they fall back on their central light. They become really perturbed only if and when minds of the geometrical type *will* force them for the time into their own quite other method. Clarity and distinctness of ideas, which might well be an index of truth in the physical sciences, were nothing of the kind—rather the opposite—in the ultimate questions of human destiny; and he was fond of

comparing the human soul's attitude to God to a dog's absolute
trust and well-founded, though incomplete, understanding of
his master.

When it came to a conflict between traditionalism and the
fully proved acquisitions of modern research, von Hügel did not
look for 'a third way', the kind of middle path down which most
Catholic iintellectuals of the day made their sometimes rather
furtive escape. He sought rather what he called 'the costing
tensions resulting from the co-existence of the two'. In fact he
was never an 'either-or' man but always for 'both-and'. There is
a touching passage in a letter of 1916, describing his return after
a long interval to Rome, to find an almost unrecognizable
improvement in the religious life of the city largely due, as he
was generously delighted to discover, to the efforts of that very
Pope Pius X whose apparent intellectual narrowness and
nervousness had been the cause of such bitter unhappiness and
injustice to himself and his scholar-friends. He had indeed a
wonderfully characteristic image for the unattractive sides of
what he called 'church appurtenance'—that institutional side of
religion, to which he attached so much importance. He
compared it in a letter to Gwendolen Greene to 'porridge or to
camphor-balls—vulgar, lumpy material, nevertheless indispensable in the one case to health, in the other to "the ethereal
camphor smell".' Or in another letter he uses the image of a
skeleton to explain what he believed to be the necessity, to any
robust and healthy religion, of a firm dogmatic substructure.
'What a hideous thing the skeleton, taken separately is, isn't it?'
he writes. 'Yet even Cleopatra, when in the splendour of her
youth, she had such a very useful, very necessary, quite
unavoidable skeleton inside her, had she not?'

Von Hügel would never have achieved the position of his later
years—which *The Times Literary Supplement* two years after
his death compared with that of Newman—if he had not struck
to the very roots of the religious problems that puzzle men and
women of all denominations and of none. Suffering, for
instance. With his typical 'costing' honesty he admitted that
Christianity had no more found an 'answer', a satisfactory
intellectual explanation of suffering, than had any other
religion. But he taught, and showed in his own person, that he
believed her to have done something more than that—to have

shown, first in her founder and then in innumerable lesser instances, how to *use* suffering, to turn what is otherwise dull, frustrating negation into positive wealth, not by Stoic endurance or listless 'resignation', but by willing acceptance. Writing to Wilfrid Ward, then in his last illness, he says:

I have long felt that it is the *apparent sterility of suffering* which adds the final touch of trial to our pains; and that this appearance is most truly *only* an appearance . . . for sin, as an offence against God, is chiefly a shirking of some effort or loneliness or pain attached to some inspiration or commandment . . . or else a seeking of some pleasure, relaxation, or vanity attached to the contrary course. And the only cure for such shirking is precisely the recovering (more and more deliberately) of that mean instinct. Pain—that real pain, which comes ready to our hand for turning into *right* pain—gets offered us by God . . .

And he goes on to suggest to the sick man to:

try more and more, at the moment itself, without any delay or evasion, without any fixed form, as simply, as spontaneously as possible, to cry out to God, to Christ Our Lord, in any way that comes most handy and the more variously the better. 'Oh! oh! This is real . . . oh may this pang deepen me, may it help to make *me* real—really humble, really loving, really ready to live or die with my soul in Thy hands.

And it was typical of him that after a serious operation not long before his death he wrote of 'three weeks in a nursing-home with really *solid* suffering the first fortnight of the time'—as though this at last meant that the time had not been wasted.

On the other hand, if he subscribed whole-heartedly to the traditional motto 'no cross, no crown' he never ceased insisting on the primary duty for the Christian of joy. It was the absence of this essential joy that he found so dispiriting and disappointing in Cardinal Newman; and he was absolutely emphatic that the purely physical explanation was inadequate. No one was readier than von Hügel to make generous allowances for the large role played by the body in hampering or de-forming the progress of the soul: he had all too much evidence from his own personal experience. But he contrasted with Newman his old friend and master, the Abbé Huvelin, who spent thirty-five years as a supernumerary curate, in an unpaid capacity in a large parish in central Paris:

There he was, suffering from gout in the eyes and brain and usually lying prone in a darkened room; he served souls with the supreme authority of self-oblivious love and radiated spiritual joy and expansion, brought light and purity and heart to countless troubled, sorrowing, or sinful souls.

Among these souls of course was Charles de Foucauld, as well as the Baron himself. To obtain and preserve this profound joy, he believed, it was absolutely essential for the soul to possess and cultivate two levels and kinds of action and interest—what he called in his curiously apt, if archaic English 'a disporting-ground in relief of the dreariness and strain of our directly religious life in times of desolation or dryness'. 'God,' he wrote, 'is overflowing Love, Joy, and Delectation'; but joy means the cultivation of the man-God relationship not only in its purest, exclusively religious sense but over the whole range of the created world. And this inclusiveness, the deep-lying optimistic faith that it implies, he believed to be of the very essence of Catholicism. As he wrote to Algar Thorold: 'The Church, hurrah! remains unchangeably greater than any and every spirituality of one direct dimension only. Catholicism is no Pietism, however sublime.'

Although under an official cloud for most of his mature life and less appreciated by what he touchingly called 'my own people' than by Anglicans and Non-conformists, von Hügel appears now—some forty years after his death—as a prophet of almost all the recent changes of attitude in the Church whose official policies caused him so much agony of mind during his lifetime. His advice to those who approached him about becoming Roman Catholics was so cautious as to be almost cautionary, particularly if his questioner was a practising member of any Christian communion. His easy and generous interchange of ideas and directly religious exchanges with non-Catholics, including clergy, were a model for what is only just beginning to be achieved on an official level. And all this because of his own profoundly-held belief that there is a large spark of the divine in all the great world religions, whose points of agreement are far more important and reach far deeper than their differences: the same belief that lies at the bottom of the present ecumenical movement.

At the basis of von Hügel's most cherished belief lay the hope that Catholics in general—and eventually, presumably, the official representatives of the Church themselves—would come to distinguish plainly between what is of the essence of the faith and what is of only secondary, temporary significance and therefore subject to change. This very distinction is at the moment occasioning what may prove to be one of the largest, as it were, submarine earthquakes in the Church's history and one which may result in a massive change of mental and psychological contours within the next two generations. Von Hügel would be loud now, as he was in his lifetime, in insisting that the 'men of little faith' are the timid and the conservative, who too easily believe an article of faith to be identical with its pictorial expression and instinctively fear that they may lose heaven itself if they relinquish its location 'above the bright blue sky'. Von Hügel on the other hand was aware how deeply inadequate, approximate and, as it were, metaphorical must be all our statements about God, all our definitions of His essence. Writing to Marcel Hébert, whose own religious beliefs were taking on an increasingly 'symbolist' character which finally led him outside the Church, he spoke of the Personality of God in terms which reflect very well his approach to the whole subject and his own personal character:

To purify, enlarge, spiritualize our own character and, especially, thus, our conception of human personality; to awaken, and keep ever more awake, our sense of the necessary inadequacy of every idea we can form of God, the absolute Spirit. But in the end to apply it (the conception of personality, that is) with these two conditions, as *what we best understand to be the best* and to be *what God cannot but be*—in a degree and way inconceivable to us.

28

SÖREN KIERKEGAARD[*]

'Sören Kierkegaard, the deep, melancholy, strenuous, utterly uncompromising Danish religionist, is a spiritual brother of the great Frenchman Blaise Pascal,' said a great nineteenth-century thinker, whose German training enabled him to sum up even this complex character in two sentences. 'In his ever occasional yet intense, diffuse yet over-concentrated, one-sided yet magnificently spiritual writings, we are given admirably fresh experiences and warnings.' Those antitheses are absolutely right, for Kierkegaard was consciously an antithetical phenomenon, a 'sign of contradiction'. The root experience of his existence was one of 'not belonging'. In the complacently bourgeois, provincial world of mid-nineteenth-century Copenhagen he was always aware of being an *extraordinarius*, 'a single letter printed backwards in a line of type', 'a rainbird', 'a corrective'.

With a brilliant intelligence, he inherited from his father a melancholy, introspective temperament which coloured all his thinking; and the realization of his father's weakness and fallibility—something that comes as a painful shock to most boys—was in his case traumatic, since it coincided with a religious crisis in his own life. In fact it was not so much the mysterious sense of guilt for the disorders or 'sins', that followed as a direct result of this shock, which made him break off—against all his natural instincts—his engagement to Regine Olsen. It was rather the growing awareness of his strange religious vocation: a kind of roving commission as 'a spy in heaven's service', as he put it; and the gradual realization that he could not saddle an unsuspecting wife with a companion called to such a life—a life of acute suffering. He was within a few years of his early death (at the age of forty-one) when he entered

* The *Ampleforth Journal:* Autumn 1967.

in his Journal without further comment that 'to love God and to be loved by him is to suffer'.

In that Journal, which Kierkegaard kept from 1834 (when he was twenty-one) until within a few months of his death just over twenty years later, almost every page carries some reference to these sufferings: speculations on their origin and nature, meditations on their significance, communications to himself on how to bear and to use them. He was from the first aware of their psychosomatic nature—indeed he says explicitly that their root lay 'in a false relationship between soul and body'. An apparently general reflection on the acute sense of shame and inhibition that civilization itself, and more particularly 'spiritualization', can bring into sexual relationships—'especially on the man's side'—may well be autobiographical. Certainly the imagery that he uses to describe the nature of his sufferings is revealing. 'The body is a heavy, sweat-soaked poultice that the soul longs to tear off,' he says. 'Like a steamer in which the engine is too powerful for the hull . . . so do I suffer.' There is an unmistakably personal note of resentment, too, in his observation of the animal-minded majority's contempt for the modesty and bashfulness that he regarded as inseparable from spirituality—in this a true child of his Puritan upbringing.

Kierkegaard was never for a moment in any doubt about his own extraordinary gifts. At one time he even described his 'thorn in the flesh' (or the painful imbalance of spirit and body in his make-up) as 'the price exacted by heaven for a spiritual power *unique among my contemporaries*'. Their inability to recognize this he could never forgive. 'Denmark stands self-condemned,' he wrote, for this failure. And although in society he would make fun of his position—the typical self-protecting gambit of a proud and sensitive temperament—yet he rebukes himself for doing so: and in the year before his death he compared his fate to the slow and ludicrous torture of being trampled to death by geese or to the hideous end of being smeared with honey and eaten by insects. 'The honey is my fame,' he adds sourly.

Yet suffering, that almost to the end of his life Kierkegaard could not accept from his fellow-men without resentment, either open or masked, he learned not only to accept but even to welcome when he recognized it as coming from the God in his relationship to Whom there was so much that was unmistakably filial—an extension and sublimation of his relationship to his father:

If You seem slow to help, it is not slowness but wisdom; if You seem slow to help, it is not slowness but because you know the speed of Your help; and if You seem slow to help, it is no paltry holding-back, but a fatherly thrift—saving up what is best for a child in the safest place and for the most suitable moment.

His 'thorn in the flesh' he came to regard as a kind of 'orthopaedic splint' necessary for a man who would live the life of the spirit. He was well aware of what his temperament could have made him without the confining, directing limitations of his suffering—proud, bitter, superior, and contemptuous of the average human existence. Of course, when Kierkegaard writes of his sins and failings, we must always bear in mind that it is no ordinary, everyday Christian writing, but a man in love with the ideal of perfection that he has seen in Christ and agonizingly aware of his failure to approximate to it. The crude sins of his very short salad days as a student even he recognized as 'not perhaps so very dreadful in God's sight'. But they were quite enough to give to so sensitive a mind an understanding of the nature of sin that he never forgot. There is a deep truth, he says, in the old myth according to which the man who once entered the Venusberg could never find his way back. Because although sins are indeed forgiven to the sincerely penitent man, he must still retrace his way—painfully, step by step—to the point where he left the narrow path. This return journey he describes vividly in the terms of a war-ravaged landscape. He is addressing in imagination a young man, begging him to mend his ways:

. . . so that you may never know the sufferings of one who has wasted the strength of his youth in rebellion against God and must then, weak and exhausted, begin the long journey back, through countries that have been laid waste and provinces stripped bare by conflicting armies, among towns destroyed by the flames and the smoking ruins of disappointed hopes, trampled harvests, and shattered authority—a journey as long as a farmer's 'bad year', as long as eternity.

Does anyone describe so graphically a journey that he has not, in some sense, himself made? Certainly Kierkegaard was very clear that he owed nothing but gratitude to the apparent cruelty of Providence in giving him his 'thorn in the flesh'. ' "Periissem nisi periissem" is my motto,' he says. 'I should have been lost (in the real, spiritual sense) if I had not been lost (in the superficial, worldly sense).' At the end, he realized that it was precisely his sufferings that gave meaning to his life, because they were an essential condition of loving God and being loved by Him. As a child, he says, he prayed for worldly success, living on terms of easy intimacy with God, as with a father whom he could confidently ask for material things. Then, as his understanding deepened, he came to believe that he might achieve a closer intimacy with God by suffering, though it seemed presumptuous to pray for such suffering. But finally 'one realizes that He is after all too infinite for us' and prayer has become 'a quiet abandoning of everything into His hands, because I am not quite sure how I *should* pray'. Did Kierkegaard know, one wonders, that he was simply following the classical pattern of all the masters of the spiritual life?

There are occasional moments in the Journal when Kierkegaard, for all his intellectual brilliance, seems to range himself with the 'holy idiots'—St Joseph of Cupertino and the Russian *yurodivy*—the clowns of the spiritual life. 'I am a Janus,' he says, 'one face laughs and the other face weeps'; and he compared his impotent desires and designs to the passions of a eunuch—once again the 'odd man out', the *extraordinarius par excellence*. Even the irony which was his chief weapon and which he wore like a kind of intellectual dandy, was an abnormality—'like the liver of a Strasbourg goose', as he put it. Yet for intellectuals themselves he had very little use; and he quotes with delight Lichtenberg's observation that 'talking to an intellectual packed with literary and historical information but devoid of any real ideas of his own is like reading a cookery book when one is hungry'. The objections to Christianity, which are generally treated as intellectual, come—he says—far more frequently from moral insubordination. The only absolute certainty is the ethical-religious: 'How blessed it is to have faith—yes, and the more blessed the higher the price that one has paid for it: just as a lover delights in winning his wife by some great sacrifice.' The sole index of faith is love; and love is the works of love, not an emotional experience of any kind whatever.

During the last years of his life, Kierkegaard's anti-intellectual bias becomes markedly stronger in the Journal. He is bitterly intolerant of what he mocks as 'the tyranny of the microscope', describes the invention of wireless telegraphy as the triumph of the Lie and believes that 'all corruption will in the end come from the natural sciences'. This impatience was, of course, not so much with genuine scientists as with the vulgarizers of science—and particularly with the journalists, for whom Kierkegaard (having suffered at their hands) always had a quite special dislike and distrust. (I particularly like his idea of a society of 'total abstainers from newspaper-reading' as something far more harmful than brandy-drinking.) The claims of the intellect seemed to him nowhere more exaggerated and debatable then in the religious sphere:

The highest thing is not to *understand* the highest, but to *do* it. . . . Have you ever seen a boat grounded in the mud? It is impossible to free it because there is no firm purchase for a pole. In the same way our whole generation is grounded in the mud of the intellect—and not disturbed by the fact, but full of the complacency and conceit that always accompany the intellect and its sins. Oh! how much easier it is to cure the sins of the heart or the flesh than the sins of the intellect!

One of his chief complaints against the Danish Lutheran Church of his day was that it had 'substituted lecturers and professors for saints and ascetics', so that Christianity in Denmark was presented as an intellectual system instead of as a way of life. In fact the last months during which he kept the Journal are filled with highly uncomplimentary remarks about Luther himself: his consecration of mediocrity, his substituting the 'public' for the Pope, his narrowness of vision ('a patient excellent at describing his symptoms,' says Kierkegaard, 'but quite without the general view needed by the doctor who can cure the disease'), his elevation into a norm of what was in fact a 'corrective', and finally his failure to become a martyr. This was a role in which Kierkegaard for a time saw himself, imagining (not without reason) the Danish clergy incensed by his attacks and urging on the mob, who already believed him mad, to an act of violence that should cost him his life—an anticipation, in fact, of the Manolios of Kazantzakis's *Christ Recrucified*. But this, he decided, would be to involve the ordinary man in a gross injustice, and he contented himself with the cry of 'back to the cloister from which Luther broke away!' It was not monastic asceticism, he says, that ruined the Middle Ages but the triumph of worldliness in the religious orders themselves. No wonder poor Regine was puzzled and told him that 'he would be sure to end by becoming a Jesuit'. Kierkegaard himself, on the other hand, saw a great similarity between the permissive, mediocre spirituality of Danish nineteenth-century religion and that preached by the seventeenth-century French Jesuits, so castigated by Pascal. The 'chemin de velours' set out in Father Le Maine's *Dévotion aisée* of 1652 was much the same, he says, as that recommended by many Danish preachers—whom he compares to castrati singers for the seductive sweetness of their tone and their lack of virile power.

Reading the Journal one often has the impression that Kierkegaard—like his contemporary Newman—would have preferred a certain amount of superstition in his countrymen to the arid enlightenment and ignoble self-seeking which he saw around him. He was particularly interested by the superstitious nature of many geniuses (one thinks at once of Schoenberg and his superstitions about numbers and dates) and of many criminals; and he explains this by the fact that both are often exploring new, unknown and perhaps in some sense 'forbidden' territories, where the traditional half-truths and probabilities, by which most men conduct their lives, cannot be relied upon. Kierkegaard even seems to have anticipated, with extraordinary clairvoyance, a characteristic of mid-twentieth-century anti-rationalism—the desire of the individual to lose himself in some form of communal intoxication, which may be political (as in the German National Socialist rallies), religious (as in the various revivalist sects), or simply animal-emotional as in drug-taking and sexual promiscuity. He compares this craving to that shown in the scenes on the Blocksberg in Goethe's *Faust:*

The desire to lose oneself, as it were to volatilize one's identity and raise it to a higher power in a state of literal, 'ec-stasy', where one no longer knows what one is doing or saying, or what force it is that is speaking through one, as the blood pulses more urgently in the veins, eyes glitter and stare and the passions come to boiling point.

No man as intellectually gifted as Kierkegaard could possibly have approved such an abdication of rationality; and in fact it

was only the intellect's claim to exclusive rights that he wished
to challenge:

A rich man with lights on his carriage sees the road rather better than the
poor man in the dark. But his carriage lights prevent him seeing the stars as
the poor man does. And so it is with all secular intelligence: it improves the
sight of things near at hand but robs its possessor of the vision of eternity.

Instead of a purely intellectual yardstick, Kierkegaard imagines
two tests by which the quality of a man can be judged. The first
is the size of the gap between his understanding and his will. 'A
man,' he says, 'should be able to compel his will to follow his
intelligence; it is between understanding and willing that the
excuses lie, and the tergiversations.' The second index of a
man's quality can be found in his instinctive willingness to serve.
The rather superior person easily commands the obedience of
his fellow men; but the absolutely superior man is by definition
a servant, because his only relationship to lesser men is a
religious one: both he and they are children of the same Father;
and he is the older, stronger brother.

No wonder, then, if the Journal contains some impatient
comments on the Olympian Goethe, whose relationship with his
'inferiors' was not at all that of either a servant or a brother.
Kierkegaard complains of the crude male egoism revealed in
such women's characters as Klärchen and Gretchen (the
diminutives themselves are revealing) and finds Goethe in
Dichtung und Wahrheit no more than 'a gifted defender of
commonplaces. . . able to talk himself out of anything—girls,
the idea of love, Christianity'. Culture can only too easily make
people insignificant—perfect them as 'copies' but destroy their
individuality. And he instances the common, popular names of
flowers and birds compared with their academic nomenclature.
This is, of course, no more than the poet's preference for poetry
to science; and Kierkegaard was quite consciously a poet.

In order that it may one day be possible even to speak of
Christianity returning to Denmark, he says in one place, 'a
poet's heart must first break—and I am that poet'. The Journal
is full of images and metaphors that amply justify this claim.
God's dealings with men, seen from too close (as we perforce see
them) are like lace, which under the microscope looks clumsy
and lacking in design. Or he compares the pleasure of

swimming—stripping naked and plunging into a foreign element that one has learned to negotiate—to that of speaking a foreign language, another way of stripping off one's everyday self. Human nature he compares to 'a skilled marksman's arrow, aimed at the mark and never resting until it reaches it: so man was made by God, for God and can never rest' (echoes of St Augustine) 'until he is in God.'

In July 1838, when Kierkegaard was twenty-five, he wrote in his Journal:

I want to work to achieve a far more intimate relationship to Christianity. Hitherto I have witnessed to the truth, as it were, from outside. I have carried Christ's cross purely externally, like Simon of Cyrene.

The seventeen years covered by the remainder of the Journal are in a sense no more than a chronicle of that interiorizing, deepening, and intensifying of his religious faith. In 1852 we find him pouring out his heart in gratitude to God for having educated him slowly and mercifully, demanding in return only complete trust and complete honesty. In the end even his sufferings have become a source of joy, simply because they are seen to be inseparable from being a tool in God's hand. 'In unseeing faith to consent to become nothing, a mere tool or vehicle', as he puts it, in writing of the Blessed Virgin. The 'sword that pierced her heart', he suggests, was not simply the agony of a mother watching her son tortured to death, but the agony of doubt—had it all been a delusion, her vocation and her son's mission? It was because Kierkegaard himself knew very well such moments of doubt that he could write that 'properly understood, every man who truthfully desires a close relationship to God, and to live in His sight has only one task—always to be joyful.' Not happy—which is something not within a man's control and altogether more superficial—but filled with the conviction of carrying out the purpose for which one came into existence. That is the root-nature of Joy; and this conviction speaks increasingly clearly in Kierkegaard's Journal as his health declines and his outward circumstances deteriorate.

The memory of his broken engagement, though it haunted him all his life, never made him bitter or contemptuous about marriage. If he felt himself called, as he put it, to 'a more decisive existence' than marriage, he appealed for justification to the

New Testament; and it was always his wish to preach to, and to be understood by, simple, ordinary people—so that it was in his eyes one of the worst crimes of the journalists, that they had prejudiced simple people against him by making him out to be mad. That he was neither a madman, a sea-green intellectual, a reactionary thinker, or a mere illuminist can be seen from a magnificent passage about goodness and freedom:

> The greatest thing that can be done for any creature is to make it free . . . and perfect freedom can only be bestowed by one who is himself perfectly free. Only God's omnipotence can withdraw at the same time as it gives itself . . . and so God's omnipotence is His goodness. For goodness is to give oneself absolutely, but in such a way that one gradually withdraws and makes the recipient truly free.

Kierkegaard was perfectly clear in his own mind that he could never win recognition in his lifetime, though he was confident of being recognized after his death. If people began to recognize me now, he says, I should have to prevent it 'by new mystifications'. For it was his task to speak in riddles, 'to come up against the wind, like thunder', to stand himself as a question-mark, not 'within quotation marks': 'As a writer I am a genius of a strange kind—subject to no authority myself, and so perpetually on my guard against becoming an "authority" to anyone else.' To read this Journal is not only to meet a 'holy hypochondriac'—a phrase that Kierkegaard himself enjoyed—but to watch the spectacle of 'that flight of the wild bird over the heads of the tame' that genius always presents.

INDEX